Quantum Legacies

QUANTUM LEGACIES

Dispatches from an Uncertain World

DAVID KAISER

With a Foreword by Alan Lightman

The University of Chicago Press

Chicago and London

The University of Chicago Press, Chicago 60637
The University of Chicago Press, Ltd., London
© 2020 by The University of Chicago
Published 2020
Printed in the United States of America

29 28 27 26 25 24 23 22 21 20   1 2 3 4 5

ISBN-13: 978-0-226-69805-2 (cloth)
ISBN-13: 978-0-226-69819-9 (e-book)
DOI: https://doi.org/10.7208/chicago/9780226698199.001.0001

Library of Congress Cataloging-in-Publication Data

Names: Kaiser, David, author.
Title: Quantum legacies : dispatches from an uncertain world /
David Kaiser ; with a foreword by Alan Lightman.
Description: Chicago : University of Chicago Press, 2020. |
Includes bibliographical references and index.
Identifiers: LCCN 2019037201 | ISBN 9780226698052 (cloth) |
ISBN 9780226698199 (ebook)
Subjects: LCSH: Quantum theory. | Physics—History.
Classification: LCC QC173.98.K358 2020 | DDC 530.12—dc23
LC record available at https://lccn.loc.gov/2019037201

♾ This paper meets the requirements of ANSI/NISO Z39.48–1992
(Permanence of Paper).

For Ellery and Toby:
quantum wondertwins

CONTENTS

FOREWORD

:: Alan Lightman

Physics has always attracted those thinkers who might have been philosophers, or musicians. Newton. Einstein. Heisenberg.

Autumn. There's a smell of fall leaves through an open window. We sit at a long table. A baker's dozen of us, I would guess. An associate provost of the institution, a couple of theater directors, a couple of playwrights, a financial person or two, an actress constantly changing her face as if putting on makeup in the dressing room, and a few physicists. David Kaiser at one end of the table, always razor sharp in dress and attentiveness. Another physicist enters the room and apologizes for being late. We are discussing the sponsorship of new plays about science, some not even yet written, to be performed at a theater nearby. It is a happy collision between science and art, between the icy equations of Newton and Einstein and the heart-wrenching dramas of Chekhov and Wilde. The deliberate joined with the spontaneous. The rational with the intuitive. The quantitative with that which cannot be quantified. Several times a year, we gather here to enter and bless this world between worlds.

It isn't surprising that Professor Kaiser perches at one end of the table. Trained as both a physicist and a historian of science, over the years he has developed into a first-rate storyteller, with his articles and books tracing not only scientific ideas but also the personal lives of the scientists and the institutional and cultural forces shaping the landscape. All revealed in the delightful essays in this book. But still. Why so many physicists at the table?

Physics attempts to reduce the world to a small number of fundamental particles and forces—the bare minimum to explain everything in the physical cosmos. I would argue that such a project reflects a deep human desire to make order out of the world, both the animate and the inanimate, and thus far transcends the mathematical runes of Newton and Einstein. The Gestalt psychologists say that we humans unavoidably try to reduce the world into meaningful patterns. Perhaps that act is what holds off insanity. The constellations, for example. If we see a random collection of dots, we parse it into some kind of figure against a background. If we see a broken circle, we mentally fill in the missing pieces. Henry Adams tried to understand the historical rise and fall of human civilizations in terms of energy and the second law of thermodynamics. Plato claimed that God made the world out of just four fundamental elements: fire, earth, air, and water. Ancient India used only three: fire was associated with bone and speech, water with blood and urine, earth with flesh and mind. We are pattern seekers and pattern makers. Physics is the ultimate distillation of that need. Art as well. Perhaps that is why there are so many physicists at this table.

Although these various human needs work together within our psyche, modern science owes much of its success to envisioning a disembodied, dispassionate universe out there, to walling itself off from the personal, and particularly the emotional. The scientists who planned the first human landing on the moon in 1969 calculated trajectories and rocket thrusts so that the spacecraft and the moon would be at the same place at the same time. Those calculations did not include the mood of the astronauts or anything about their personal lives.

A closely related issue: science does not normally concern itself with the narrative of its own (human) history of discovery. In this sense, as well as in others, science distinguishes itself from the humanities. The body of knowledge in the humanities might be called "horizontal"—the seminal works of all eras are considered equally relevant and enriching. The works of no era are considered superior to those of any other era. When we study philosophy, for example, we might begin with Confucius and Plato and Aristotle and move toward Nietzsche and Bertrand Russell. Or with literature, we might begin with the *Iliad* and the *Odyssey* and slowly advance to *The Great Gatsby*. There is no sense in which *The Great Gatsby* is considered more "correct" or wiser than the *Iliad*. By contrast, science is a vertical endeavor. The theories, data, and knowledge of each century are thought to improve and replace the understanding of previous centuries. In predicting the orbits of planets and other gravitational phenomena, Einstein's theory of gravity is simply more accurate than Newton's theory of gravity. Period. A graduate student in science studies the most up-to-date results of her field before launching into research at the "frontiers" of the field. There is no time, and often little interest, in the history of the subject, which now lies in the dusty stacks of libraries alongside candles for lighting and slide rules for calculating. My college textbook on heat, titled *Thermal Physics*, is full of equations describing the modern understanding of heat as the *random motion* of atoms and molecules. There isn't a single word about the earlier theory of heat as a material fluid, called phlogiston. Or about the colorful story of Benjamin Thompson (1753–1814), who began life as a schoolteacher in New Hampshire and later fled to England after the fall of Boston in the Revolutionary War and became head of the Bavarian army. It was Thompson who discovered that heat was motion, not substance, by way of his job superintending the boring of cannon in the military arsenal at Munich. In his essay read before the Royal Society of London in 1798, Thompson wrote, "It frequently happens that in the ordinary affairs and occupations of life, opportunities present themselves of contemplating some of the most curious operations of Nature. . . . I was struck with the very considerable degree of Heat which a brass gun acquires, in a short time, in being bored . . . in the Friction of two metallic surfaces."

In contrast to the usual view of science as a disembodied enterprise, the essays in this book are full of the human drama and historical context that accompany scientific discovery and knowledge.

The smell of the leaves and their symphony of color remind me that our human senses, as rich as they are, provide a misleading picture of the world—a picture that has been radically revised by modern physics. We've learned that, contrary to appearances, the Earth spins on its axis and races through space. We've learned that, contrary to appearances, zillions of x-rays and radio waves and gamma rays stream by us each second, invisible to the eye. We've learned that in the tiny world of the atom, particles behave as if they could be in several places at once and, further, seem to have instantaneous effects on each other, violating the usual notions of cause and effect. These last discoveries are the subject matter of quantum physics, a central theme of Kaiser's new book. If we must question the validity of our sense perceptions, if we must give up our intuitive understanding of "reality," if we must accept a new narrative of the physical world, let us at least do so with the charming human stories in this book.

Introduction

"Don't laugh," physicist Paul Ehrenfest had scribbled on a tiny scrap of paper. He was attending a scientific meeting in Brussels with about two dozen leading physicists, late in October 1927. While his colleagues spoke, one after the other, of their struggles to make sense of the new quantum theory, Ehrenfest, like a giddy schoolboy, had passed the note to his friend Albert Einstein. A bit lower on the page, Einstein had scrawled his response. "I laugh *only* at the naiveté," he wrote back in his curvy, looping handwriting. "Who knows who will be laughing in the coming years."[1]

I still remember the jolt I felt when I came across the handwritten notes in the early 1990s. I had made a trip to the rare books library at Princeton University, my first foray into archival research. Princeton's library has an official duplicate set of Einstein's unpublished papers and correspondence, the originals of which are housed at Hebrew University in Jerusalem. The Einstein collection at Princeton consists of nearly one hundred boxes, stuffed with faded photocopies and microfilms, and I had begun poking around in a tiny corner of the vast archive. I was hardly the first person to notice the playful notes between Ehrenfest and Einstein—other historians had already quoted the exchange—but I was transfixed all the same.[2]

Holding the page in my hand, I tried to imagine the scene. Huddled in a conference room within a nondescript academic building next to the leafy Parc Léopold, Einstein, Ehrenfest, and their colleagues had scrambled to lash together a new framework with which to describe the behavior of matter at its most fundamental. In the years leading up to the meeting in Brussels, much of what previous generations of physicists had thought they knew about light and atoms seemed to unravel. Young guns at the conference, like Werner Heisenberg and Wolfgang Pauli—still in their twenties—insisted upon new ideas. Theirs was a vision of the world that was at root probabilistic, riven by inevitable trade-offs between what physicists could ever hope to know, as Heisenberg had boldly suggested just a few months earlier when he introduced his now-famous uncertainty principle. Ehrenfest and Einstein, each in their late forties, appreciated the cleverness of the new ideas but wrestled with doubts. Einstein, in particular, would soon emerge as one of the most trenchant critics of the new quantum mechanics, concerned that the theory harbored too many weaknesses to support the broad edifice of theoretical physics.[3]

Even so, as the heady debates spilled past the prepared remarks in the conference room and animated the physicists' dinners at a nearby hotel, Einstein and Ehrenfest responded to the uncertainties with humor, even humility. What a grand intellectual adventure they seemed to be en-

Figure 0.1. Paul Ehrenfest (*left*), his son Paul Ehrenfest Jr., and Albert Einstein, relaxing in the Ehrenfest home in Leiden, June 1920. (*Source*: Wikimedia Commons.)

joying, as they passed their notes back and forth, kidding their colleagues, oblivious to the darkness they would each soon face. A few years after the Brussels meeting, Hitler assumed power in Germany, forcing Einstein to flee and resettle in Princeton. The impact was even harsher for Ehrenfest, whose ebullience with friends and students had masked a creeping depression. The swirling uncertainties of worldly events exacerbated his growing unease with the rapid-fire changes in physics. "I have completely lost contact with theoretical physics," he confided at one point to a colleague. "I cannot read anything any more and feel myself incompetent to have even the most modest grasp about what makes sense in the flood of articles and books." He worried, too, about his younger son, Wassik, who suffered from Down syndrome. His letters to friends became more desperate. Five years after the Brussels meeting, Ehrenfest drafted a new note for Einstein. He wrote of his efforts, "ever more enervated and torn," to make sense of the new physics. But it all seemed too much; the collected uncertainties made him "completely 'weary of life.'" He never mailed the letter. Drafts were found late in September 1933, after Ehrenfest shot his young son, Wassik, in the waiting room of his son's physician, before turning the pistol on himself.[4]

:::

Long past the earnest and spirited debates at the 1927 Brussels meeting, quantum mechanics remains a centerpiece of physicists' description of nature. Even after all those years, physicists have yet to find a single instance in which predictions from the theory have failed to match experimental tests. And it has certainly not been for lack of trying.

A quarter century after stumbling upon Einstein's and Ehrenfest's notes, I was lucky to engage in a quantum-mechanical adventure of my own. Early in the morning of 6 January 2018, squinting in the bright sunshine, I walked across the tarmac at the airport of La Palma, a tiny member of the Canary Islands off the western coast of Morocco. At sea level, La Palma looks every bit the tropical paradise, palm trees swaying in a gentle breeze with all the postcard loveliness of Hawaii. Bleary-eyed and sleepy from my overnight flights—New York to Madrid, then several more hours to La Palma—I managed to catch up to my colleague Anton Zeilinger, a renowned physicist who has built a career designing ever more clever experiments to test the strangest properties of quantum theory. Despite his own travels that morning, Anton looked as cheery as ever. (In my experience,

Figure 0.2. On the distant hillside sits the metallic dome of the Nordic Optical Telescope, part of the Roque de los Muchachos Observatory on the Canary Island of La Palma. Visible to the left of the telescope dome is a rectangular shipping container, which served as a makeshift laboratory during our Cosmic Bell experiment in January 2018. During our brief observing time, we had to contend with freezing rain and occasional hailstorms. (*Source*: Photograph by Calvin Leung.)

Anton is unfailingly cheerful, on any continent and in any time zone.)

Anton picked up the keys to a rental car, and we drove up the mountain to the Roque de los Muchachos Observatory. Before long, the palm trees gave way to stunted brush, as we wound our way up the narrow road to an elevation of nearly eight thousand feet. As we drove up to the modest headquarters of the observatory, the picture-perfect blue sky that had greeted us at the airport had changed to a driving, freezing hailstorm.

At the observatory we met up with about a dozen members of our team, most of them young graduate students and postdocs from Anton's research group, based in Vienna. Several had been at the observatory for weeks, installing equipment and performing calibration tests. We were there to conduct a new experimental test of quantum entanglement, a phenomenon that Einstein himself had helped to identify in the years following the famous meeting in Brussels. For our new test, we were going to use two of the enormous telescopes at the observatory to collect random bits of information from the sky, gathering light from some of the furthest known quasars, all while shooting pairs of entangled particles between those telescopes from a special laser that Anton and his group had shipped over from Vienna.

After a few nail-biting nights of poor weather, the skies over La Palma finally cleared, and we were able to conduct our new experiment. Within hours we had preliminary results in hand. After a few more weeks of careful calculation, we could confirm that our latest experiment, like all previous ones, showed results perfectly in line with the quantum-mechanical predictions—and thereby in conflict with the sort of results that Einstein had thought such experiments should yield. Our experiment on the mountaintop relied upon instruments that Einstein himself never lived to see: thirteen-foot telescope mirrors polished to perfection, a high-powered laser, single-photon avalanche diode detectors, and timing circuits for our electronics disciplined to nanosecond accuracy with atomic clocks.[5] We had marshaled all these modern tools to test ideas that dated back nearly to the 1927 Brussels meeting. I couldn't help but wonder what the grand masters of quantum theory, who had gathered all those years ago for their discussions and debates, would have made of our latest efforts.

:::

Ever since my first excursion into Einstein's papers in the Princeton library, I have been riveted by a kind of doubleness of scientific research. In Einstein's day as in our own, researchers' ambition has often been to transcend the vagaries of here and now, to contribute lasting insights into how the world works that might reach beyond a given researcher's limited view. Yet each of us—today's scientists no less than Einstein and his peers—remains unavoidably embedded in a certain time and place. Scientists are immersed in the particulars of the world, moment by moment, even as many dream of superseding these accidents of history.

The particularities of time and place can shape scientific research in many ways, across many scales. At the individual level, hiccups of individual biography can produce lonely thinkers like Paul Dirac or the heroic perseverance of Stephen Hawking. Forces acting across larger scales, from changes in specific institutions to epochal geopolitical rifts, shape scientific research as well. During the years since Ehrenfest and Einstein exchanged their playful notes in Brussels not quite a century ago, the story of physics and physicists has been dramatically reshaped by these kinds of forces several times over: by the rise of Nazism and cataclysmic world war, by the relentless calculation of nuclear brinksmanship during the Cold War, and by the whiplash suddenness with which the Cold War sputtered to an end. With hindsight, we can trace how such turning points helped to propel unexpected insights about the natural world, just as often as they limited individuals' horizons.

Captivated by the ways in which the world of ideas remains tethered to more earthly concerns, I pursued graduate study in theoretical physics and in the history of science. After that, I had the great fortune to join the faculty at the Massachusetts Institute of Technology (MIT), where I have taught both subjects for twenty years. Drawing on this pair of perspectives, I have aimed in much of my writing to reconstruct the never-straight path of physics research over the past century by focusing on scientists' training. How have particular approaches come to seem natural for researchers and their students in a given time and place? How have generations of young physicists learned to broach questions and evaluate results, and how have those methods shifted — sometimes subtly, sometimes by great leaps — between one setting and another? Teaching, training, undertaking the hard work of trying to forge some shared understanding between individuals and across generations — these are the "quantum legacies" to which I refer in the book's title. Some of these legacies are captured well by focusing on the efforts of individuals and their exchanges with small circles of colleagues; others are clarified by scrutinizing new machines, like the first electronic, programmable computers or hulking particle accelerators like the Large Hadron Collider. I am particularly fascinated by textbooks as legacy-making engines: objects crafted expressly to try to smuggle forward, into the future, bundles of hard-won skills and insights. Chasing down these legacies has offered me an opportunity to reflect on my own training, as I wonder about what sorts of legacies my colleagues and I might pass along to our students.

The essays in this volume explore episodes in which physicists have grappled with subtle uncertainties of the natural world, even as they have labored within the inevitable uncertainties of the social and political worlds. Though the settings of the essays range across space and time, many center on developments within the United States during the Cold War years. Physicists' fortunes during that era lurched between periods of plenty and want. The boundless optimism coming out of the war years was countered, for some, by McCarthyist red-scare anxieties. Enormous new tools helped to reshape physicists' intellectual landscape—some of them legacy equipment from the massive wartime projects, others underwritten by a newly generous federal government, especially its military divisions. A flood of eager new students rushed into the field, making physics the fastest-growing academic specialty in American higher education. But then came crippling reversals. The first occurred in the early 1970s, years into the slog of the Vietnam War, combined with détente, "stagflation," and substantial cuts in defense and education spending. Another came in the early 1990s, after a second wave of defense-related spending, which had accelerated under the Reagan administration, vanished quickly, and unexpectedly, after the Soviet Union dissolved.

These turning points, driven by the unsteady push-pull of world events, changed the texture of everyday life for members of what had been, not long before, a rather sleepy academic field. Einstein and Ehrenfest had traded notes across the conference table at the 1927 meeting in Brussels. After the war, physicists had to adapt to rather different communication habits. That American workhorse of a journal the *Physical Review* swelled from three thousand pages published in 1951 to more than thirty thousand pages published in 1974. One of my favorite photographs shows particle physicist Val Fitch, who received the Nobel Prize in 1980, in danger of being crushed by the sheer mass of the

Figure 0.3. Particle physicist Val Fitch poses with stacks of the *Physical Review*, arranged by decade, in the late 1970s. (*Source*: Photograph by Robert P. Matthews, courtesy of AIP Emilio Segrè Visual Archives, *Physics Today* Collection.)

journal, as the stacks of each year's volumes climbed higher and higher. The journal's longtime editor Samuel Goudsmit explained to a colleague in the mid-1960s how he and his editorial team attempted to manage the changes. The journal, he wrote, "is no longer similar to the neighborhood grocery story where old customers get personal attention." Rather, it had become "more like a supermarket where the manager is hidden in an office on the top floor. As a result, lots of things are just done by routine rather than by human judgment." He meant it literally: by that time, the editorial office was experimenting with a new punch-card computer system to mechanize tasks like matching referees with submissions, tracking the progress of referee reports received, and recording responses sent to authors.[6]

The effects were palpable. Goudsmit himself noted in the mid-1950s that each issue of the journal had become "almost too bulky to carry." A few years later, he observed that "we have long ago passed the psychological limit above which the subscriber is overwhelmed by the bulk and looks only at the few articles in his own narrow field." At a time when most professional physicists in the United States still purchased their own subscriptions to the journal, several wrote to Goudsmit to complain about the runaway volume: back issues of the journal filled their office shelves and threatened to overrun closet space at home. Goudsmit advised his colleagues to stop being "overly sentimental" and simply rip out those articles they wanted from each issue, tossing the rest. "There is really little reason to keep more than about 'six feet' of *The Physical Review* at home," he concluded. Though some felt "revolted" by "such destruction of the printed word," others took up the editor's suggestion. One Caltech physicist reported with pride that he had reduced two feet of shelf space, which had been taken up by his copies of the 1963 *Physical Review*, to a few inches, though he wondered whether the journal might switch to a different glue for its binding, to make it easier to tear out the desired articles.[7]

:::

As I have glued the essays for this volume together, my thoughts have often returned to Goudsmit's advice. Various essays will no doubt resonate more strongly for some readers than others. In collecting them for this volume, I have updated most and merged others together. I have also tried to put the essays more directly into dialogue with each other by grouping them into four sections. My hope is that each essay can be enjoyed on its own, while together they may yield insights, partial and patchwork-like, into broader transitions within physicists' continuing quest to understand space, time, and matter. Like an old issue of the *Physical Review*, the picture they present is more kaleidoscopic than a formal portrait.

The essays in "Quanta" address discrete moments in the transformation of physicists' understanding of quantum theory, from the heady days of the 1920s, through the dark times of the 1930s, to some bizarre twists early in the nuclear age. The section culminates with my own group's efforts—first in Vienna and most recently at the observatory on La Palma—to build upon this century-long legacy by testing quantum entanglement as thoroughly as our imaginations and toolkits would allow.

The second section, "Calculating," focuses on some of the changes in how—and why—new generations became physicists within the United States, during and after the dramatic disruptions of the Second World War. The early years of the Cold War fostered a new type of calculation among many defense analysts and policymakers in the United States, as they scrutinized the unsteady standoff with the Soviet Union. To prepare for the nation's defense—in case the Cold War ever tipped into outright warfare between the superpowers—these analysts and policymakers concluded that the United States needed many more physicists, trained and at the ready, who would be available to staff massive projects like a next-generation Manhattan Project. The defense intellectuals' calculations, and their relentless calls for "scientific manpower," drove enormous shifts in enrollment patterns and the rhythms by which new generations entered higher education. These institutional changes, in turn, reshaped how young physicists learned to calculate and how they grappled with quantum theory.

In "Matter," I turn to physicists' more recent efforts to understand the world of electrons, quarks, and more fleeting constituents of the subatomic realm, such as the long-elusive Higgs boson. Over the past half century, physicists around the world have cobbled together a remarkably successful account of subatomic particles and the forces between them. The so-called "Standard Model" is built within the framework of quantum theory but harbors conceptual surprises that neither Einstein nor Heisenberg could have foreseen. Meanwhile, the specialty of high-energy physics inherited its own particular legacy from the Cold War: within the United States, political priorities and unprecedented federal investment fostered an era of gigantism, as physicists built larger and larger machines to probe matter at smaller and smaller scales. That investment—and the political arguments that had sustained it—collapsed soon after the Soviet Union fell apart, in the early 1990s. I was an undergraduate at the time. An internship at the Lawrence Berkeley National Laboratory provided me with a crash course in particle physics as well as in the changing political realities of "big science."

The final section, "Cosmos," explores moments in physicists' changing conceptions of space and time on the largest scales—phenomena described by that other great pillar of modern physics, relativity. Physicists' efforts to understand Einstein's relativity, and to use it to model the evolution of our universe as a whole, have paralleled their concerted efforts on quantum theory over the past century.

These efforts have yielded some stunning insights into the universe and our place within it, even as these insights have resisted physicists' every attempt to combine relativity and quantum theory into a conceptually consistent whole. Were Einstein alive today, he might be excused for passing new notes to a friend, still laughing at our naiveté.

QUANTA

1

All Quantum, No Solace

Physics became "modern" at breakneck speed. Only twenty years separated Albert Einstein's formulation of special relativity, in 1905, and the development of quantum mechanics in 1925–26. The two events have attracted rather different kinds of stories. Einstein's achievement is typically portrayed as an epic tale of one man's obsession. The creation of quantum mechanics, on the other hand, required an ensemble cast, more Heinrich Böll's *Group Portrait with Lady* (with a nod to Marie Curie) than Melville's *Moby Dick*.

And quite a cast it was. The fatherly Niels Bohr dressed like a banker and mumbled like an oracle. Werner Heisenberg, a gregarious Bavarian, thrived as the life of any party, banging out Beethoven piano sonatas into the wee hours and traipsing up mountain paths in lederhosen. Louis de Broglie, a young French aristocrat, flitted from studies of literature and history before brazenly introducing the notion that solid matter might consist of waves. Erwin Schrödinger, the dapper Austrian, led a surprisingly bohemian lifestyle. Openly promiscuous, he sustained a string of affairs with much younger women—his biographer felt compelled to list "Lolita complex" in the index—and raised, with his wife, a child he fathered with the wife of one of his assistants. Then there were the children who never grew up: practical jokesters like the brilliant Russian physicist Lev Landau and the acid-tongued Wolfgang Pauli.[1]

The creators of quantum mechanics formed a tight-knit community. When not visiting together at Bohr's Institute for Theoretical Physics in Copenhagen or at one of the informal conferences sponsored by the industrialist-turned-philanthropist Ernest Solvay, they kept up their conversations by letter. Tens of thousands of their letters have survived. Over the years, scholars have dutifully inventoried, archived, microfilmed, and translated these letters, subjecting them to the kind of line-by-line scrutiny once reserved for scripture.[2] And yet for all the attention lavished on the founding generation of quantum physicists—several biographies each of Bohr and Heisenberg, lengthy treatments of Schrödinger, Max Born, and others—few have paid much attention to the brilliant British physicist Paul Dirac. Of all the strange characters who paraded through Bohr's Copenhagen institute, Bohr called the waifish and withdrawn Dirac "the strangest man."[3]

As a young man, Dirac had dreamed of studying relativity. By the time he arrived at Cambridge in 1923 to study for a doctorate, however, the local expert on the subject had stopped taking students. Dirac was assigned to Ralph Fowler instead, at the time Britain's foremost expert on the strange physics of atoms. Quantum theory was still an ungainly patchwork of models and heuristics, and physicists across Europe had been struggling for decades to make sense of matter at the smallest scales. The tendency was to begin with the familiar laws that govern everyday objects—the motion of planets in the solar system or the interaction of electric charges and radiation—and then append this or that ad hoc rule to cover instances when the usual equations broke down.

Dirac's first glimpse of a new approach arrived by mail in September 1925. Werner Heisenberg had sent page proofs of a new article to Dirac's adviser Fowler, who in turn sent them to Dirac, still at his family home in Bristol for the summer vacation. Fowler appended the casual note, "What do you think of this? I should be glad to hear."[4] In the short article, Heisenberg aimed to establish a new quantum mechanics, a first-principles treatment of matter and radiation rather than the tattered, hand-me-down quilt of his teachers' generation. Heisenberg was convinced that it was a mistake to rely on intuitions or models taken from ordinary physics. Electrons orbiting a nucleus in an atom were not just like planets orbiting the Sun, he declared: the electrons' paths could not be observed, even in principle. The best way forward, he announced in the opening paragraphs of his brief article, written when he was just twenty-three years old, was to construct a new theory "in which only relations between observable quantities occur." In Heisenberg's new formulation, arrays of discrete numbers replaced the smoothly varying quantities usually found in physicists' equations, and he filled his arrays with observable quantities, such as the color and brightness of light emitted by atoms that had been excited by some outside source of energy.[5]

Making his way through Heisenberg's page proofs, Dirac brushed aside the opening philosophical challenge about sticking only with observable quantities. Instead, Dirac focused on something interesting later in the article. Some of the arrays of numbers in Heisenberg's new scheme behaved in a curious way: their product depended on the order in which they were multiplied. *A* times *B* did not equal *B* times *A*. Unlike Heisenberg, Dirac had a degree in pure mathematics; such unconventional rules of multiplication reminded him of similar relationships that crop up in ordinary mechanics—the physics of tops, balls, and orbiting planets—when written in a particularly advanced, mathematically elegant way. This buried mathematical analogy, rather than Heisenberg's opening salvo about unobservable quantities, prodded Dirac forward. Nine months later he completed his dissertation, using the mathematical analogy to clarify and generalize Heisenberg's work.

By then, Heisenberg's approach to quantum mechanics was no longer the only game in town. During the winter of 1926, Erwin Schrödinger—ten years older than Heisenberg and Dirac, and far more conservative in his approach to physics, if not in his personal life—had produced an independent formulation. Instead of Heisenberg's discrete arrays of numbers, so jarringly unfamiliar to most physicists, Schrödinger borrowed the familiar mathematics of waves, usually enlisted to describe such phenomena as ripples spreading on the surface of a pond or the wailing screech of a passing siren. The contrast between Heisenberg's and Schrödinger's rival approaches stirred strong emotions. In one of his early articles on wave mechanics, Schrödinger wrote that he "felt discouraged, not to say repelled," by Heisenberg's methods. Heisenberg shot back, in a letter to a friend, that the more he thought about Schrödinger's work, "the more disgusting I find it."[6]

Dirac's dissertation had earned him a scholarship to spend the 1926–27 academic year on the Continent. His

Figure 1.1. Paul Dirac (*left*) with Werner Heisenberg, early 1930s.
(*Source*: AIP Emilio Segrè Visual Archives.)

first stop was Bohr's institute in Copenhagen. There he
ignored most of the jabs and jokes of his colleagues and se-
questered himself in the library, where he set about dem-
onstrating that Heisenberg's and Schrödinger's approaches
were mathematically equivalent, the name-calling notwithstanding. Other people produced independent proofs of

the equivalence, but most physicists were quick to admire Dirac's approach as the most powerful and elegant.

Dirac now produced a steady stream of breathtaking results. Just before leaving Bohr's institute in January 1927, he extended the quantum formalism beyond atoms to the treatment of light, including the interaction of charged particles with radiation, thus creating a whole new physical theory. He dubbed it "quantum electrodynamics," or QED. Next he set about rectifying Heisenberg's and Schrödinger's equations with Einstein's special relativity, seeking to create a quantum mechanics that would hold together even as the objects under study moved at speeds closer and closer to the speed of light. Back in Cambridge in autumn 1927 (having been elected a fellow of St. John's College), he derived his relativistic equation for the electron, clarifying, along the way, the notion of quantum "spin" that had puzzled colleagues for years.

In the spring of 1931, under constant hectoring from Heisenberg and Pauli to account for a strange mathematical feature of his new equation, Dirac boldly predicted the existence of antimatter: as-yet-unseen cousins to the ordinary particles we see all around us, which have the same mass but opposite electric charge. Within two years, physicists in California and Cambridge had accumulated striking experimental evidence in support of Dirac's conjecture. Dirac thus set in motion what has become the single most precise physical theory ever. At the latest count, theoretical predictions calculated with QED match experiments to eleven decimal places. Errors on both sides of the ledger—theorists' calculations and experimenters' data—are now measured in parts per trillion.[7]

To Dirac, mathematics could be beautiful, and beauty was the surest guide to truth. "It is more important to have beauty in one's equations than to have them fit experiment," he was fond of repeating. (Today's proponents of string theory sometimes borrow his rhetoric.) A stickler for precision, Dirac developed a style that was elegant, even austere: colleagues sometimes complained that he left all the words out of his articles. Asked after a lecture to clarify an earlier point, he often responded by repeating, verbatim, what he had said the first time. He honed his mathematical notation—now universal among physicists—to allow maximum economy of expression. His style was enshrined in his famous textbook, *The Principles of Quantum Mechanics*, first published in 1930 and an instant classic. Ninety years later the book is still in print, and still much revered.[8]

Unlike some less fortunate figures in the history of science, who suffered long delays before their contributions were recognized, Dirac skyrocketed to the top of the profession. He was elected to the Royal Society at the unheard-of age of twenty-seven. (In a further rarity, he was elected upon his first nomination.) In July 1932, just shy of his thirtieth birthday, Dirac was elevated to the Lucasian Chair of Mathematics at Cambridge—once held by the equally precocious Isaac Newton and later occupied by Stephen Hawking. He shared the 1933 Nobel Prize with Schrödinger and remains one of the youngest recipients. Though he was by no means finished as a physicist by that point—he continued to produce important, intriguing work in quantum theory and cosmology, most of which would bear fruit only later in the hands of others—the five-year period following his dissertation was surely one of the most brilliant and far-reaching bursts of scientific creativity ever recorded.

In other respects, Dirac was a late bloomer. He seems to have become interested in politics only in his early thirties, when he developed a fascination with the "Soviet experi-

Figure 1.2. Physicists gather at the Institute for Theoretical Physics in Copenhagen for a 1933 conference. *Front row, from left to right*: Niels Bohr, Paul Dirac, Werner Heisenberg, Paul Ehrenfest, Max Delbruck, and Lise Meitner. (*Source*: Nordisk Pressefoto, courtesy of AIP Emilio Segrè Visual Archives, Margrethe Bohr Collection.)

ment." He made several trips to the Soviet Union in the 1930s to collaborate with leading physicists. He stunned guests at the banquet in honor of his Nobel Prize by delivering an impromptu harangue on the importance of protecting workers' wages in the midst of the global economic crisis.

He resisted getting involved in the war effort after 1939, even as many of his colleagues in physics and mathematics rallied to the cause. He performed calculations for the "Tube Alloys" project—Britain's early nuclear weapons program—and consulted on at least one occasion with Klaus Fuchs, the German-born British physicist who later spied for the Soviets from deep within the Manhattan Project. But when J. Robert Oppenheimer, the scientific director of the Los Alamos laboratory, asked Dirac to work full-time on the Manhattan Project, he declined.

Dirac's leftist sympathies posed some problems after the war. He was denied a visa to enter the United States in April 1954, at the height of American anti-Communist hysteria. (This public rebuke came just as Oppenheimer was undergoing his own withering interrogation before the US Atomic Energy Commission's personnel security board, although Oppenheimer's case was still secret at the time.)[9] Nearly two decades later, as Dirac made plans to settle in the United States for his retirement, he was barred from accepting professorships at some universities because of his long-standing membership in the Soviet Academy of Sciences.

Yet these public setbacks and embarrassments were minor compared with Dirac's private struggles, as captured in Graham Farmelo's moving biography, *The Strangest Man* (2009). Indeed, as Farmelo's book makes clear, Dirac's success is all the more remarkable given the travails of his personal life. Six months before he was sent Heisenberg's page proofs in the autumn of 1925, his older brother, Felix, committed suicide. They had both studied engineering at the eminently practical Merchant Venturer's College in Bristol, the engineering faculty of Bristol University; their father taught French in the adjoining secondary school.[10] Year after year, Felix had watched his younger brother gallop past him scholastically. Their father paid scant attention to Felix and refused to support his interest in medicine. To the end of his life, Dirac blamed his father for Felix's death.

The Dirac household was crushingly joyless. Paul's cold and authoritarian father spent decades in an open feud with his mother. Drawing on a stash of family correspondence, Farmelo documents the efforts made by Dirac's grasping mother to receive some solace from her younger son. She leaned on him for emotional support of a kind that few people in their teens or early twenties could be expected to provide, and this seems only to have pushed Dirac deeper into his shell. In a country that has long celebrated individuals of quiet reserve, Dirac became a man of almost inhuman reticence. (In Cambridge, a "Dirac" was a unit of measurement corresponding to one word per hour, the "smallest imaginable number of words that someone with the power of speech could utter in company," as Farmelo puts it.)[11] Dirac explained late in life that his father had forced him to speak only French at the dinner table, and frustrated by his lack of facility with the language, he had found it easiest to remain silent. Meals became so stressful for the young Dirac that he developed severe indigestion; he was afflicted by food sensitivity throughout his life.

Family dynamics like these, combined with Dirac's famously peculiar demeanor, understandably invite some form of psychological speculation. Retrospective diagnosis has become something of a pastime in our psychologizing age. Abraham Lincoln wasn't just solemn or moody, several scholars have argued; he suffered from clinical depression. Isaac Newton, whose father died before he was born and whose mother abandoned him at age three upon her remarriage, acted out a lifetime of hostile priority claims for want of secure childhood attachment—or so Frank Manuel concluded in his 1968 biography, *A Portrait of Isaac Newton*.[12] (Nor are these diagnoses limited to historical figures. My wife, a psychologist, has no patience for the great Russian novels. According to her, Raskolnikov could have been set right with a modest regimen of psychotropic drugs; *Crime and Punishment* should have been a jaunty five-page pamphlet, with a happy ending to boot.)

Farmelo treads into similar territory in the closing pages of his book. He marches through several traits commonly associated with autism—sensitivity to food and to loud, sudden noises; extreme reticence and awkward social behavior; obsessive focus on a few arcane topics—adding each time that the appearance of similar traits in Dirac was probably not coincidental. In the end, he can't resist it: "I believe it to be all but certain that Dirac's behavioural traits as a person with autism were crucial to his success as a theoretical physicist."[13]

Such claims demand a leap of faith. There is, of course, the basic challenge of evidence: even with the rich, textured family correspondence that Farmelo mines so well, traces left behind in letters seem rather different from hours of professional observation or the directed interrogation of a trained clinician. More important is a tacit ontological assumption that today's repertoire of psychiatric diagnoses transcends time and place. Does the "melancholia" so often described by Lincoln's contemporaries really map smoothly onto today's "clinical depression"? The American Psychiatric Association changes entries in its *Diagnostic and Statistical Manual of Mental Disorders*—the industry standard for cataloging psychiatric ailments—often dramatically, on a timescale of decades. Yesterday's diseases can become today's idiosyncrasies; one era's nervous disorders are another's quirky tics.[14] Why, then, this need to pathologize genius? Is it, perhaps, to let ourselves off the hook? No wonder we failed to achieve the greatness of a Newton, Lincoln, or Dirac, we may well comfort ourselves. They weren't just smarter than us; their brains were different from ours.

Whether or not we follow Farmelo on this brief excursion into diagnosis-at-a-distance depends on how closely we hew to Heisenberg's dictum to focus only on observable quantities. After all, neither "melancholia" nor "autism" seems to correspond, in any straightforward way, to states that historians can measure. (Surely it is telling that just four years after Farmelo's book appeared, the editors of the *Diagnostic and Statistical Manual* replaced "autism" with "autism spectrum disorder," with a consequent shift in definitions and diagnoses.)[15]

Diagnoses aside, Farmelo's book suggests a different, rather remarkable way in which Dirac's particular mind-set left its mark on the theory he worked so hard to advance. Nearly a century after its formulation, quantum theory remains scientists' most successful and precise description of nature ever devised. Yet it is a curiously minimalist description, forcing physicists to choose, for example, between saying something definite about where a particular particle is at a given moment and where it is going—they cannot say both. Dirac's strict economy of expression and his famous reticence have shaped how generations of physicists discuss the quantum world. For all its astounding success, the account that we glean from quantum mechanics—matter behaving at times in familiar ways, only to surprise us, suddenly, with its inescapable oddity—remains fitfully strange. Not unlike Paul Dirac himself.

2

Life-and-Death

When Nature Refuses to Select

Of all the bizarre facets of quantum theory, few seem stranger than those captured by Erwin Schrödinger's famous fable about the cat that is neither alive nor dead. It describes a cat locked inside a windowless box, along with some radioactive material. If the radioactive material happens to decay, then a device will detect the decay and release a hammer, which will smash a vial of poison and kill the cat. If no radioactivity is detected, the cat will live. Schrödinger dreamed up this gruesome scenario to criticize what he considered a ludicrous feature of quantum theory. According to proponents of the theory, before anyone opened the box to check on the cat, the cat was neither alive nor dead; it existed in a strange, quintessentially quantum state of alive-and-dead.

Today, in our LOLcatz-saturated world, Schrödinger's strange little tale is often played for laughs, with a tone more zany than somber.[1] It has also become the standard-bearer for a host of quandaries in physics and philosophy. In Schrödinger's own time, Niels Bohr and Werner Heisenberg proclaimed that hybrid states like the one the cat was supposed to be in were a fundamental feature of nature. Others, like Einstein, insisted that nature must choose: alive or dead, but not both.

Although Schrödinger's cat flourishes as a meme to this day, discussions tend to overlook one key dimension of the fable: the environment in which Schrödinger conceived of it in the first place. It's no coincidence that, in the face of a looming world war, genocide, and the dismantling of German intellectual life, Schrödinger's thoughts turned to poison, death, and destruction. Schrödinger's cat, then, should remind us of more than the beguiling strangeness of quantum mechanics. It also reminds us that scientists are, like the rest of us, humans who feel—and fear.

:::

Schrödinger crafted his cat scenario during the summer of 1935, in close dialogue with Albert Einstein. The two had solidified their friendship in the late 1920s, when they were both living in Berlin. By that time, Einstein's theory of relativity had catapulted him to worldwide fame. His schedule became punctuated with earthly concerns—League of Nations committee meetings, stumping for Zionist causes—alongside his scientific pursuits. Schrödinger, originally from Vienna, had been elevated to a professorship at the University of Berlin in 1927, just one year after introducing his wave equation for quantum mechanics (now known simply as the "Schrödinger equation"). Together they enjoyed raucous Viennese sausage parties—the *Wiener Würstelabende* bashes that Schrödinger hosted at his house—and sailing on the lake near Einstein's summer home.[2]

Figure 2.1. Erwin Schrödinger relaxes with a pipe and a drink. (*Source*: Photograph by Wolfgang Pfaundler, courtesy of AIP Emilio Segrè Visual Archives.)

Too soon, their good-natured gatherings came to a halt. Hitler assumed the chancellorship of Germany in January 1933. At the time, Einstein was visiting colleagues in Pasadena, California. While he was away, Nazis raided his Berlin apartment and summer house and froze his bank account. Einstein resigned from the Prussian Academy of Sciences and quickly made arrangements to settle in Princeton, New Jersey, as one of the first members of the brand-new Institute for Advanced Study.[3]

Meanwhile, Schrödinger—who was not Jewish and had kept a lower profile, politically, than Einstein—watched in horror that spring as the Nazis staged massive book-burning rallies and extended race-based restrictions to university instructors. Schrödinger accepted a fellowship at the University of Oxford and left Berlin that summer. (He later settled in Dublin.) In August, he wrote to Einstein from the road, "Unfortunately (like most of us) I have not had enough nervous peace in recent months to work seriously at anything."[4]

Before too long their exchanges picked up again, their once-leisurely strolls now replaced by transatlantic post. Prior to the dramatic disruptions of 1933, both physicists had made enormous contributions to quantum theory; indeed, both earned their Nobel Prizes for work on the subject. Yet both had grown disillusioned with their colleagues' efforts to make sense of the equations. Danish physicist Niels Bohr, for example, insisted that according to quantum theory, particles do not have definite values for various properties until they are measured—as if a person had no particular weight until stepping on her bathroom scale. Moreover, quantum theory seemed to provide only probabilities for various events, rather than the rock-solid predictions that flowed from Newton's laws or Einstein's relativity. Bohr's arguments failed to sway Einstein or Schrödinger. Now separated by an ocean but armed with paper and postage stamps, they dove back into their intense discussions.[5]

In May 1935, Einstein published a paper with two younger colleagues at the Institute for Advanced Study, Boris Podolsky and Nathan Rosen, charging that quantum mechanics was incomplete. They contended that there exist "elements of reality" associated with objects in the world—properties of physical objects that had definite values—for which quantum theory provided only probabilities.[6] In early June Schrödinger wrote to congratulate his friend on the latest paper, lauding Einstein for having "publicly called the dogmatic quantum mechanics to account over those things that we used to discuss so much in Berlin." Ten days later Einstein responded, venting to Schrödinger that "the epistemology-soaked orgy ought to come to an end"—an "orgy" they each associated with Niels Bohr and his younger acolytes like Werner Heisenberg, who argued that quantum mechanics completely described a nature that was, itself, probabilistic.[7]

This exchange produced the first stirrings of the soon-to-be-born cat. In a follow-up letter to Schrödinger, Einstein asked his friend to imagine a ball that had been placed in one of two identical, closed boxes. Prior to opening either box, the probability of finding the ball in the first box would be 50 percent. "Is this a complete description?" Einstein asked. "NO: A complete statement is: the ball is (or is not) in the first box."[8] Einstein believed just as fervently that a proper theory of the atomic domain should be able to calculate a definite value. Calculating only probabilities, to Einstein, meant stopping short.

Encouraged by Schrödinger's enthusiastic reply, Einstein pushed his ball-in-box analogy even further. What if the small-scale processes that physicists were used to talking about were amplified to human sizes? Writing to Schrödinger in early August, Einstein laid out a new scenario: imagine a charge of gunpowder that was intrinsically unstable, as likely as not to explode over the course of a year. "In principle this can quite easily be represented quantum-mechanically," he wrote. Whereas solutions to Schrödinger's own equation might look sensible at early times, "after the course of a year this is no longer the case at all. Rather, the ψ-function"—the wave function that Schrödinger had introduced into quantum theory back in 1926—"then describes a sort of blend of not-yet and of already-exploded systems." Not even Bohr, Einstein crowed in his letter, should accept such nonsense, for "in reality there is just no intermediary between exploded and not-exploded."[9] Nature must choose between such alternatives, Einstein insisted, and so, therefore, should the physicist.

Einstein could have reached for many different examples of large-scale effects with which to criticize a quantum-probabilistic description. His particular choice—the unmistakable damage caused by exploding caches of gunpowder—likely reflected the worsening situation in Europe. As early as April 1933, he had written to another colleague to describe his view of how "pathological demagogues" like Hitler had come to power, pausing to note that "I am sure you know how firmly convinced I am of the causality of all events"—quantum and political alike. Later that year he lectured to a packed auditorium in London about "the stark lightning flashes of these tempestuous times." To a different colleague he observed with horror that "the Germans are secretly re-arming on a large scale. Factories are running day and night (airplanes, light bombs, tanks, and heavy ordnance)"—so many explosive charges ready to explode. In 1935, around the time of his spirited exchange with Schrödinger about quantum theory, Einstein publicly renounced his own prior commitment to pacifism.[10]

Perhaps inspired by their latest exchange, Schrödinger began writing a long essay of his own, on "the present situation in quantum mechanics." A week and a half after receiving Einstein's letter about the exploding gunpowder, Schrödinger replied with a novel twist. In place of gunpowder, there was now a cat.

"Confined in a steel chamber is a Geiger counter prepared with a tiny amount of uranium," Schrödinger wrote to

Figure 2.2. Nazi book-burning rally on the Opernplatz in Berlin, 1933. (*Source*: Imagno, courtesy of Getty Images.)

his friend, "so small that in the next hour it is just as probable to expect one atomic decay as none. An amplified relay provides that the first atomic decay shatters a small bottle of prussic acid. This and—cruelly—a cat is also trapped in the steel chamber." Just as in Einstein's example, Schrödinger imagined the appointed time elapsing. Then, according to quantum mechanics, "the living and dead cat are smeared out in equal measure." Einstein was delighted. "Your cat shows that we are in complete agreement," he wrote in early September. "A ψ-function that contains the living as well as the dead cat just cannot be taken as a description of the real state of affairs."[11]

A few months after Einstein's September letter, Schrödinger's now-famous cat example appeared, with nearly identical wording, in the magazine *Die Naturwissen-schaften*.[12] But it almost didn't make it into print. Days after he submitted his draft to the magazine, the founding editor—a Jewish physicist named Arnold Berliner—was fired. Schrödinger thought about retracting the essay in protest and relented only after Berliner himself interceded.[13]

Schrödinger's thoughts that summer were preoccupied with more than just concerns about Berliner's mistreatment. Schrödinger had made no secret of his distaste for the Nazi regime and had become downright fatalistic when forced to flee Berlin, musing in his diary, "might it not be the case that I have already learnt enough of *this* world. And that I am prepared . . ." Months after arriving in Oxford, a visiting friend noted how unhappy he was, the pressures of displacement compounding the effects of the dismal, daily news. In May 1935—just as the Einstein, Podolsky, and Rosen paper appeared in print—Schrödinger delivered a twenty-minute lecture on BBC radio entitled "Equality and Relativity of Freedom," recalling the many times throughout history in which "gallows and stake, sword and cannons have served to free respectable people" from political repression.[14] Against the drumbeat of advancing fascism, little wonder that talk of balls in boxes morphed so quickly into explosions, poisons, and morbid calculations of life and death.

While his essay was in press, Schrödinger wrote to Bohr, trying again to discern how Bohr and the others could make peace with the bizarre features of quantum mechanics. As with Einstein, Schrödinger longed to discuss such matters with Bohr in person, "but the times are now little suited for pleasure trips." Larger questions loomed. Schrödinger wrote of his "wish once again to be somewhere permanently, that is, to know with considerable probability what one is to do for the next 5 or 10 years."[15] Living only with probabilities had taken its toll.

Yet Europe sank deeper into darkness. Just a few years after Schrödinger introduced his fable about the quantum cat and prussic acid, Nazi engineers began using the self-same poison—under the trademarked name "Zyklon B"—in their brutally efficient gas chambers. In March 1942, just before his scheduled deportation to a concentration camp, Schrödinger's former editor from *Die Naturwissenschaften*, Arnold Berliner, killed himself—choosing, in the end, a terrible certainty.[16]

:::

In time, the challenge that Schrödinger thought would undercut quantum mechanics became, instead, one of the most familiar tropes for teaching students about the theory. A central tenet of quantum mechanics is that particles can exist in "superposition" states, partaking of two opposite properties simultaneously. Whereas we often face "either-or" decisions in our everyday lives, nature—at least as described by quantum theory—can adopt "both-and."

Over the decades, physicists have managed to create all manner of Schrödinger-cat states in the laboratory, coaxing microscopic bits of matter into "both-and" superpositions and probing their properties. Despite Schrödinger's reservations, every single test has been consistent with the predictions from quantum mechanics. As I take up in the next chapter, for example, my colleagues and I at MIT recently demonstrated that neutrinos—subatomic particles that interact very weakly with ordinary matter—can travel hundreds of miles in such catlike states.[17]

There is a double irony, then, to Schrödinger's tale of his twice-fated cat. First, although Schrödinger's cat remains well known within (and beyond) physics classrooms, few recall that Schrödinger introduced his fable to criticize quantum mechanics rather than elucidate it. Second, and even more telling: Schrödinger's cat served, in its day, as synecdoche for a broader world that had become too strange—and, at times, too threatening—to understand.

3

Operation: Neutrino

Every fraction of a second, invisible particles called neutrinos whiz past the vans and Winnebagos on Highway 169
headed toward McKinley Park in northeastern Minnesota,
just shy of the Canadian border. Having begun their journey at Fermilab, an immense physics laboratory outside
Chicago, some of those speeding neutrinos smack into five-
ton slabs of steel within an underground mine in the town
of Soudan, Minnesota (population: 446), sending sparks of
charged particles arcing toward sensitive detectors. Quite
unlike the camper vans and RVs, the neutrinos complete
their journey—450 miles across the Upper Midwest—in
less than three-thousandths of a second.

Neutrinos are fundamental to the construction of the
universe. They are tremendously abundant, outnumbering
atoms by about a billion to one. They modulate the reactions
that cause massive stars to explode as supernovas. Their
properties provide clues about the laws governing particle

physics. And yet neutrinos are among the most enigmatic particles, largely owing to their reticent nature: they have no electric charge and practically no mass, so they interact extremely weakly with ordinary matter. Some sixty-five billion of them stream through every square centimeter of your body—an area the size of a thumbnail—every second, without your ever noticing them.

Through elaborate sleuthing, physicists have identified three distinct types of neutrinos, which differ in their subtle interactions with other particles. Stranger still, the neutrinos can "oscillate" between types, shedding one identity and adopting another as they travel through space. That discovery led to a significant expansion of the standard theory of how particles behave. Still more recently, my colleagues and I studied neutrinos' subtle oscillations to probe an even deeper mystery of matter.

We used data on the detection of neutrinos in the Soudan mine to complete one of the longest-distance tests of quantum mechanics ever conducted. In particular, we demonstrated that the tiny neutrinos make the journey in a "superposition" state—miniature versions of Schrödinger's fabled cat. Along the way, the neutrinos are in no definite state, but rather a quintessentially quantum-mechanical hybrid of all three of the neutrino types that physicists have identified. Only when the neutrinos are measured at the Soudan mine, all those miles away from Fermilab, do they snap into one definite state or another, just as Schrödinger's cat assumes a definite condition—dead or alive, but not both—once an observer opens the box to look.

In scarcely more than half a century, then, neutrinos have gone from wispy, exotic particles at the edge of detectability to tools for investigating matter at its most essential—from prizeworthy quarry to something more like a forensics kit. In retracing that transformation, we catch glimpses of a larger story, of physicists groping toward an abstruse, beguiling account of nature, set against (and at times engulfed by) larger dramas of the nuclear age.

:::

The discovery of the neutrino dates back to the 1930s, when the Italian physicist Enrico Fermi helped hammer out the first workable theory of nuclear phenomena like radioactive decay. To make his calculations work—in essence, as a bookkeeping device to ensure that all the energy entering a nuclear reaction would balance all the energy at the end—Fermi's colleague Wolfgang Pauli had postulated the existence of a new, undetected particle that carried some of the energy away. Fermi developed the idea more fully and dubbed the mystery particle a neutrino, or "little neutral one," since it was theorized to carry no electric charge.[1]

Neither Fermi nor anyone else at the time thought that such tiny wisps of matter could ever be detected directly. Before long, the spread of fascism in Europe quickly overshadowed such questions. As nations mobilized for war, physicists on all sides of the conflict were absorbed into top-secret projects. Meanwhile, the introduction of Nazi-inspired racial laws in Italy put Fermi's family in danger (Fermi's wife, Laura, was Jewish). Late in 1938, he managed a *Sound of Music*–like escape, exploiting a trip to Stockholm to accept the Nobel Prize in order to slip out of Europe and head for the United States, where he became one of the early scientific leaders of the Manhattan Project. In December 1942, Fermi's group in Chicago coaxed the first nuclear reactor to go critical, inducing controlled nuclear fission. Their reactor design was scaled up during the war to produce plutonium for atomic bombs.[2]

By the time peacetime research resumed, physics had undergone an enormous transformation. The bloodiest armed conflict in history had thundered to an end, punctuated by the use of nuclear weapons against the cities of Hiroshima and Nagasaki. Throughout much of the war, the hastily built laboratory at Los Alamos, New Mexico, had served as the main coordinating site for the Manhattan Project.[3] After the war the laboratory continued to focus on improving and expanding the nation's nuclear arsenal, giving physicists a newfound level of prestige—and funding. In this new environment, the first serious effort to detect neutrinos began at Los Alamos, in the early 1950s.

Frederick Reines was a young physicist at the Los Alamos Laboratory, part of the team that tested new weapons at the Eniwetok Atoll in the middle of the Pacific Ocean. Returning from a series of bomb tests in late spring 1951, he fell into a discussion about neutrinos with Fermi, who was then visiting the lab. Reines realized that the aboveground nuclear blasts that he and his team were studying at places like Eniwetok should produce enormous floods of neutrinos—so many that a wayward few just might be detectable.[4]

Reines and another Los Alamos colleague, Clyde Cowan, convinced the laboratory director to let them conduct an experiment at an upcoming bomb test. They would first excavate a narrow, deep hole near where the bomb would be detonated. Inside the shaft they would suspend a one-ton detector, so big that they nicknamed it "El Monstro." When the bomb went off, carefully sequenced electronics would release the detector, letting it fall freely while the enormous shock wave from the bomb rumbled through the surrounding earth. (If they had bolted their detector in place so close to the blast, the shock wave would have ripped it apart.) Moments later, after the shock wave had passed, the detector would land on a pile of feathers and foam rubber.

Buried at the bottom of the shaft, the device would be awash in neutrinos from the nuclear fireball. Sensitive electronics on the detector—a huge vat filled with a solution of toluene, an organic molecule usually found in paint thinners—would monitor for telltale flashes of light. A flash would indicate that one out of hundreds of trillions of neutrinos had struck some matter in the liquid and shaken loose a positron, the antimatter twin of the electron. Meanwhile, the physicists would have to wait several days, until the hazardous radioactivity on the surface had died down sufficiently, before they could return to ground zero, dig back down the 150-foot shaft, and retrieve their instrument.[5]

While preparing for their bomb-based test, Reines and Cowan realized that they could also hunt for neutrinos in a less dramatic way. With one more tweak to their protocol to better rule out spurious readings, they could install their liquid-filled detector next to a nuclear reactor rather than beneath an exploding bomb. The two researchers ran a pilot test near one of the huge reactors in Hanford, Washington—supersized versions of Fermi's original reactor. Satisfied with the results, they installed an improved device at the newer, more powerful reactor at Savannah River, in South Carolina, during the fall of 1955. (The Savannah River plant was built to produce tritium for hydrogen bombs, which are thousands of times more destructive than the original nuclear bombs.) Within months, Reines and Cowan had clocked in enough tiny flashes of light to convince their colleagues—and, in time, the Nobel Prize committee—that they had finally caught sight of the elusive neutrino.[6]

Figure 3.1. Frederick Reines (*left*) and Clyde Cowan conduct an early test to try to detect neutrinos, using reactors in Hanford, Washington, in 1953. (*Source*: Courtesy of the University of California–Irvine.)

: : :

Fermi's former assistant Bruno Pontecorvo followed these developments with particular interest. Pontecorvo had begun his career in the 1930s as the youngest member of Fermi's group in Rome. (Older members nicknamed him the "Puppy.") He immersed himself in the esoterica of nuclear physics, including Fermi's new theory of radioactivity and the still-emerging, and still-hypothetical, ideas about neutrinos. Coming from a large Jewish family, Pontecorvo eventually found his situation in Italy untenable. His flight from fascism was even more dramatic than Fermi's, closer to *Casablanca* than *The Sound of Music*: first a fellowship to study in Paris, then a harrowing, nighttime escape through the countryside in June 1940, as Nazi tanks rolled into the city. From the south of France, Pontecorvo boarded a train to Madrid and then on to Lisbon before catching a steamship to New York City.[7]

Once in North America, Pontecorvo, too, began working on the Manhattan Project. He was assigned to one of the British contingents working in Montreal, aiming to perfect a different type of nuclear reactor than Fermi's in Chicago. After the war, he took a position with the new British nuclear research facility in Harwell, near Oxford, to continue his research on reactors. Around that time, he proposed a scheme to try to detect neutrinos streaming out of a nuclear reactor, years before Reines and Cowan independently pursued the idea.

Two fascinating books—Simone Turchetti's *The Pontecorvo Affair* (2012) and Frank Close's *Half-Life* (2015)—document the next bizarre twists in Pontecorvo's life story. He had been named as one of the inventors, together with Fermi and other members of the Rome group, on a patented technique to slow down certain nuclear particles and thereby increase the rate of particular nuclear reactions. The technique proved to be pivotal to wartime studies of nuclear fission, both for reactors and for bombs. After the war, the patents, which had been granted in Italy in 1935 and in the United States in 1940, assumed a radically different patina.[8]

In 1949, other members of the Rome group sought compensation for their patented technique, which had since been built directly into the sprawling infrastructure of the US nuclear complex. The patent-wrangling triggered an FBI investigation, which dredged up long-forgotten material on several of Pontecorvo's relatives who had been outspoken Communists in Italy. A few weeks later, one of Pontecorvo's colleagues at Harwell, Klaus Fuchs, confessed to having passed atomic secrets to the Soviets during the war. Like Pontecorvo, Fuchs was an émigré from the Continent who had served as part of the British delegation on the Manhattan Project. The associations suddenly looked suspicious.[9]

What came next reads like a Le Carré novel. While vacationing in Italy in early September 1950, Pontecorvo and his family abruptly zigzagged from Rome to Munich to Stockholm, then on to Helsinki, where they met up with Soviet agents. Pontecorvo's wife and young children got into one car while Pontecorvo climbed into the trunk of another, and the secret caravan drove through the forest into Soviet territory. Hours later the group arrived in Leningrad; within days they had been delivered to Moscow. Weeks went by before the British or American authorities caught on. Finally, the US Joint Congressional Committee on Atomic Energy released a thick report, *Soviet Atomic Espionage*, proclaiming Pontecorvo's defection to be only slightly less damaging than Fuchs's disloyalty, and even worse than the alleged espionage for which Ethel and Julius Rosenberg would later be executed.[10]

While lurid headlines swirled in Britain and the United States, Pontecorvo settled into a new position at the vast nuclear research facility at Dubna, outside Moscow. As Close reveals, on the basis of his scrutiny of Pontecorvo's personal notebooks from the time, Pontecorvo consulted on aspects of the secret Soviet nuclear weapons project, at least for a while. But soon he was given space to pursue more basic re-

Figure 3.2. Bruno Pontecorvo walks through the streets of Moscow in March 1955, after defecting to the Soviet Union with his family. (*Source*: Photograph by Hulton Archive, courtesy of Getty Images.)

search as well. Upon learning of Reines's and Cowan's discovery, his thoughts returned to his long-standing love, the neutrino.[11]

In 1957, Pontecorvo published an article in the leading Soviet physics journal suggesting that neutrinos could oscillate between different varieties, or "flavors." (Translations of the journal into English had recently begun, secretly underwritten in part by the Central Intelligence Agency.)[12] He refined this idea in a series of papers, reasoning from quantum theory that the neutrinos should occur in superposition states, neither in one flavor state nor the other but both. (He considered only two types of neutrinos at the time.) Whenever physicists made a measurement, they should find a given neutrino in one flavor state only. But in between observations the neutrinos would not have a fixed identity. They would live in a state of statistical indeterminacy, partly one flavor and partly the other.[13]

How stark the contrast must have seemed between the quantum world and the rules of human affairs. Fluid, uncertain identities had no easy place at a time when McCarthy-era investigators pressed, "Are you now or have you ever been . . . ?"[14] Yet Pontecorvo himself had ricocheted among several distinct identities in rapid order, from the young "Puppy" of Fermi's group in Rome to "Academician Bruno Maximovitch Pontecorvo" for the KGB.

One of the earliest dividends of Pontecorvo's theory concerned physicists' understanding of the Sun. The core of the Sun is a massive nuclear reactor, and physicists could exploit their theories of nuclear physics to predict to high precision how many neutrinos from the Sun should be detected on Earth. Yet more sensitive follow-up experiments to the original Reines-Cowan test had found only about one-third the expected number of solar neutrinos. Amid the early stirrings of détente between the United States and the Soviet Union in the late 1960s, Pontecorvo was able to share his latest ideas directly with colleagues in the West. He now calculated that neutrinos should oscillate among three distinct flavors. If so, the solar neutrino detectors, which were sensitive to only one of the flavors, should register about the number of neutrinos that the experimentalists kept finding. Years' more data confirmed the pattern and eventually convinced the skeptics.[15]

The solar neutrino readings provided only indirect evidence that neutrinos oscillate. The next challenge was to try to catch them in the act. Groups around the world built ever-bigger detectors buried deep underground, ultimately thousands of times larger than Reines and Cowan's original design. During the late 1990s and early 2000s, teams at the Super Kamiokande facility in Japan, and separately at the Sudbury Neutrino Observatory (SNO) in Ontario, Canada, amassed compelling evidence of neutrino oscillations. The existence of oscillations indicated that neutrinos could not be massless particles, as was predicted by the prevailing theory of the time. The origin and nature of neutrino mass remains a major, ongoing area of exploration in physics. Physicists also continue to test whether only three flavors of neutrinos exist in nature. Any more than three would provide decisive evidence that physicists' Standard Model of particle physics—which has successfully described every experiment involving elementary particles for more than forty years—is incomplete.

The SNO and Super Kamiokande projects netted the Nobel Prize for physicists Arthur McDonald and Takaaki Kajita, the leaders of the two groups, in October 2015. Three weeks later, the annual Breakthrough Prize in Fundamental Physics disbursed $3 million among the nearly 1,400 physicists who had worked on the teams.[16] My friend and colleague at MIT, Joseph Formaggio, a member of the SNO collaboration, used his share of the prize money to buy a nice bottle of wine—something a few price points north of his usual purchases.

:::

Research on neutrinos seems more exciting than ever, offering a tantalizing route to press beyond the Standard Model. My own interest in them, however, was sparked when Joe suggested that we might put neutrinos to work in a different way: to test one of the central tenets of quantum theory. Pontecorvo suggested back in the 1950s that neutrinos' flavor-changing ways were directly analogous to Schrödinger's half-dead/half-alive cat. If so, then neutrino oscillations could provide a powerful way to explore the validity of superposition. Joe realized that we could analyze how the mix of neutrino flavors changes as the particles travel, finally settling into a single flavor when measured. Together with two marvelous students, undergraduate Talia Weiss and graduate student Mykola Murskyj, Joe and I set out to investigate.

Pontecorvo's theory of neutrino oscillations, based squarely on the notion of quantum superpositions, provides an excellent match to the latest experimental data. But we wondered: could the same data be compatible with alternate theories? Perhaps a theory more like the type that Einstein and Schrödinger had held out hope for—in which superpositions are absent and particles always possess definite properties at each instant of time—could account equally well for the data. Joe's key insight was that if neutrinos are truly governed by quantum superposition—if they zoom through space as "both-and" rather than "either-or"—then the likelihood of measuring particular flavors at a detector should be quantitatively different than if each neutrino possessed a definite identity at any given moment and merely oscillated among distinct identities over time.

Though our analysis became a bit baroque, in essence it boiled down to a simple observation. According to quantum mechanics, the probability of detecting a particular flavor of neutrino spreads out through space and time like a wave. The wave associated with one neutrino flavor evolves with a slightly different frequency than the wave for another flavor. For a neutrino in a superposition state, those not-quite-identical waves can interfere with each other, in much the way that overlapping waves can interfere with each other on the surface of a pond. At some points along the neutrino's journey, the crests of each probability wave will align, while at others the trough from one wave will cancel out a crest from another.[17]

All of this leads to a measurable effect. Where crests meet, the probability of detecting a particular flavor rises; where troughs cancel crests, that probability falls. Moreover, the interference pattern—those spots where crests add with crests—should shift with the neutrino's energy. On the other hand, in rival theories that lack superposition, such as those sought by Einstein and Schrödinger, no such interference pattern should occur. We calculated the different patterns predicted for the number of neutrinos that should be detected in a given flavor as one varied their energy, depending on whether the neutrinos made their journey in a superposition state or not. Then we compared these calculations with data from the Main Injector Neutrino Oscillation Search, or MINOS, an experiment that had been shooting beams of neutrinos from Fermilab toward the Soudan mine in Minnesota since 2005.

Not only did the quantum-mechanical calculation match the MINOS data beautifully, but the Einstein-like version didn't come close. Even taking into account the uncertainties and statistical flukes that can skew experimental results, we found the odds that the neutrinos were genuinely governed by an Einstein-like theory of matter, with no superpositions, to be less than one in a billion.[18]

Whereas quantum effects like superposition are usually manifest only over incredibly short distances of tens or hundreds of nanometers, our test demonstrated unmistakable quantum strangeness over a distance of 450 miles. And that may be just the beginning. After all, we are awash in neutrinos from the Sun, and cutting-edge experiments, like the IceCube Neutrino Observatory at the South Pole, can now detect primordial neutrinos that have been traveling through space for billions of years, ever since the big bang. Perhaps neutrinos like these, which have traversed cosmic distances, can also be coaxed to reveal telltale signs of quantum superposition. Then we could test this central feature of quantum theory across the vastness of space itself.

In the meantime, by puzzling through the strange dance of oscillating neutrinos, my colleagues and I have found that for all the fairy-tale strangeness of quantum theory, its predictions hold up across human-sized distances. Perhaps it is fitting that the neutrinos' journey from Fermilab to the Soudan mine is about the same distance that Pontecorvo himself traveled during his storied lifetime, bounding from Rome to Paris or sneaking from Helsinki to Moscow. Across distances like these, we can say with confidence, the world really is governed by quantum superpositions.

4

Quantum Theory by Starlight

The headquarters of the National Bank of Austria, in central Vienna, are exceptionally secure. During the week, in the basement of the building, employees perform quality-control tests on huge stacks of euros. One night in April 2016, however, part of the bank was given over to a different sort of testing. A group of young physicists, with temporary ID badges and sensitive electronics in tow, were allowed up to the top floor, where they assembled a pair of telescopes. One they aimed skyward, at a distant star in the Milky Way. The other they pointed toward the city, searching for a laser beam shot from a rooftop several blocks away. For all the astronomical equipment, though, their real quarry was a good deal smaller. They were there to conduct a new test of quantum theory.

It is difficult to overstate the weirdness of quantum physics. Even Albert Einstein and Erwin Schrödinger, both major architects of the theory, ultimately found it too outlandish to be wholly true. For one thing, unlike Newtonian physics and Einstein's relativity, which elegantly explained the behavior of everything from the fall of apples to the motion of galaxies, quantum theory offered only probabilities for various outcomes, not rock-solid predictions. Einstein objected that quantum theory treated objects in the real world as mere puffs of possibility—both there and not there or, in the case of Schrödinger's famous imaginary cat, both alive and dead. Strangest of all was what Schrödinger dubbed "entanglement." In certain situations, the equations of quantum theory implied that one subatomic particle's behavior was bound up with another's, even if the second particle was across the room or on the other side of the planet or in the Andromeda galaxy. They couldn't be communicating, exactly, since the effect seemed to be instantaneous, and Einstein had already demonstrated that nothing could travel faster than light. In a letter to a friend, Einstein dismissed entanglement as "spooky actions at a distance"— more ghost story than respectable science.[1] But how to account for the equations?

Physicists often invoke twins when trying to articulate the more fantastical elements of their theories. Einstein's relativity, for instance, introduced the so-called twin paradox, which illustrates how a rapid journey through space and time can make one person age more slowly than her twin. (Schrödinger's interest in twins was rather less academic, centering on his exploits with the Junger sisters, who were half his age.)[2] I am a physicist, and my wife and I actually have twins, so I find it particularly helpful to think about them when trying to parse the strange dance of entanglement.

:::

Let us call our quantum twins Ellie and Toby. Imagine that, at the same instant, Ellie walks into a restaurant in Cambridge, Massachusetts, and Toby walks into a restaurant in Cambridge, England. They ponder the menus, make their selections, and enjoy their meals. Afterward, their waiters come by to offer dessert. Ellie is given the choice between a brownie and a cookie. She has no real preference, being a fan of both, so she chooses one seemingly at random. Toby, who shares his sister's catholic attitude toward sweets, does the same. Both siblings like their restaurants so much that they return the following week. This time, when their meals are over, the waiters offer ice cream or frozen yogurt. Again the twins are delighted—so many great options!—and again they choose at random.[3]

In the ensuing months, Ellie and Toby return to the restaurants often, alternating aimlessly between cookies or brownies and ice cream or frozen yogurt. But when they get together for Thanksgiving, looking rather plumper than last year, they compare notes and find a striking pattern in their selections. It turns out that when both the American and British waiters offered baked goods, the twins usually ordered the same thing—a brownie or a cookie for each. When the offers were different, Toby tended to order ice cream when Ellie ordered brownies, and vice versa. For some reason, though, when they were both offered frozen desserts, they tended to make opposite selections—ice cream for one, frozen yogurt for the other. Toby's chances of ordering ice cream seemed to depend on what Ellie ordered, an ocean away. Spooky, indeed.

Einstein believed that particles have definite properties of their own, independent of what we choose to measure. (He famously pressed a colleague, while strolling at night through Princeton, whether the colleague really believed that the Moon was only in the sky when someone happened to look.)[4] Einstein believed with equal fervor that local actions can produce only local effects. In describing our quantum twins, in other words, Einstein would have insisted that Toby had some definite dessert preference on his own each evening, regardless of what type of dessert Ellie's waiter happened to offer. After all, given that no information can travel faster than light, it seemed self-evident that Ellie's order should have no bearing on Toby's actions, once the twins had traveled sufficiently far apart. If relativity really set an absolute speed limit on how quickly A could influence B, then Toby would need to carry all his own information with him as he traveled to his restaurant; there would be no time to receive an update on what his dessert order should be based on whatever had just happened with Ellie.

In 1964, the Irish physicist John Bell identified the statistical threshold between Einstein's world and the quantum world.[5] If Einstein was right, then the outcomes of measurements on pairs of particles should line up only so often; there should be a strict limit to how frequently Toby's and Ellie's dessert orders are correlated. But according to quantum theory, Bell went on to show, the correlations should occur significantly more often. For the past four decades, scientists have tested the boundaries of Bell's theorem. In place of Ellie and Toby, they have used specially prepared pairs of particles, such as photons of light. In place of friendly waiters recording dessert orders, they have used instruments that can measure some physical property, such as polarization—whether a photon's electric field oscillates along or at right angles to some direction in space. To date, every single published test has been consistent with quantum theory.[6]

From the start, however, physicists have recognized that their experiments are subject to various loopholes, circumstances that could, in principle, account for the observed results even if quantum theory was wrong and entanglement merely a chimera. One loophole, known as locality, concerns information flow: could a particle on one side of the experiment, or the instrument measuring it, have sent some kind of message to the other side before the second measurement was completed? Another loophole concerns statistics: what if the particles that were measured somehow represented a biased sample, a few spooky dessert orders amid thousands of unseen boring ones? Physicists have found clever ways of closing one or the other of these loopholes over the years, and beginning in 2015, several beautiful experiments have managed to close both at once.[7]

But there is a third major loophole, one that Bell overlooked in his original analysis. Known as the freedom-of-choice loophole, it concerns whether some event in the past could have nudged or previewed the choice of measurements to be performed and thereby affected the behavior of the entangled particles—in our analogy, the desserts being offered and the selections that Ellie and Toby made. If the twins knew ahead of time the exact order in which, say, Toby would be offered baked goods or frozen goods, then they could have devised a plan so that their dessert orders would betray certain patterns. (As Schrödinger himself commented in 1935, one would hardly be surprised if a student aced an exam if he had received a copy of the questions ahead of time.)[8] Where the locality loophole imagines Ellie and Toby, or their waiters, communicating with each other while various desserts are being offered in each restaurant, the freedom-of-choice loophole supposes that some third party could have either guessed in advance what one of the waiters would offer or — stranger still — somehow forced the waiter's hand. It was this third loophole that my colleagues and I recently set out to address.

:::

I began thinking about the freedom-of-choice loophole in the autumn of 2012. I had recently finished a book on the early history of Bell's theorem and the first efforts to test quantum entanglement in the laboratory, in an era before the topic had entered physicists' mainstream.[9] With the book done, I began to work with a new postdoctoral researcher at MIT, Andy Friedman; our plan had been to focus together on various theoretical models of the early universe, trying to account for the behavior of our cosmos around the time of the big bang. Just as he was getting settled at MIT, Andy had dinner in Harvard Square with a friend of his from graduate school, Jason Gallicchio. Jason's new office was about to be a good bit further away: he had begun working with the South Pole Telescope collaboration and would soon deploy to Antarctica to serve as station science leader for the "winterover" shift. (Astronomers like to joke about the winterover position: you need to work only one night, although, at the pole, that night happens to last six months.)

Before leaving Cambridge for Antarctica, Jason had been thinking about the great vastness of space, and all that astronomers and cosmologists had learned in recent decades about the structure of spacetime. In human terms, the speed of light is enormous — nearly seven hundred million miles per hour — and yet our universe is so large, and has been expanding for so long, that some faint pinpricks of light in the night sky, which astronomers have carefully measured and cataloged, hail from objects so far away that the light has been traveling, uninterrupted, for most of the history of the universe.

Jason and Andy mused that night over burgers: Could we somehow exploit these large-scale features of the universe to test quantum theory? What if we made astronomical measurements of the light from very distant objects and used the outcomes of those observations to determine which measurements to perform, here on Earth, on pairs of entangled particles? In that case, rather than flipping a coin in the kitchen to determine which dessert option to offer, Ellie's and Toby's waiters would offer choices of dessert based on events that had occurred long ago and far away. Whereas a coin in the kitchen might have been tampered with by some hidden mechanism right up to the moment that the waiters took Ellie's and Toby's orders, the astronomical signals would hail from opposite sides of the universe.

Knowing of my interest in John Bell's work and in cosmology, Andy shared with me what he and Jason had brainstormed together, and soon all three of us set to work— Andy now ensconced at MIT just down the hall from my office, and Jason from his new perch at the bottom of the world.[10] Our biggest stroke of luck came just as we were finishing our paper proposing the new protocol for experimental tests of Bell's inequality. Seemingly by chance—or perhaps thanks to the subtle gears of entanglement—Austrian physicist Anton Zeilinger was scheduled to visit MIT and deliver a lecture for the physics department. Over a remarkable career, Zeilinger had designed and conducted ever more ingenious experiments to test some of the strangest, most

Figure 4.1. The Cosmic Bell collaboration takes form, nourished by a working lunch near MIT in October 2014. *From left to right*: Andrew Friedman, Jason Gallicchio, Anton Zeilinger, and David Kaiser. (*Source*: Photograph from the author's collection.)

beguiling features of quantum theory, including Bell's inequality.[11] I immediately secured time on Zeilinger's schedule during his visit. At our appointed hour, Andy and I pitched to him our idea of using uncorrelated, astronomical sources of randomness for Bell tests. Midway through our spiel, Anton sported a grin as wide as Andy's and mine. He and his group in Vienna had recently completed a significant project on aspects of the freedom-of-choice loophole, and he seemed tickled by our novel twist.[12] Before long, we had put together a team: our "Cosmic Bell" collaboration was born.

:::

We performed our first experiment in April 2016, spread out in three locations across Schrödinger's native Vienna. A laser in Zeilinger's laboratory at the Institute for Quantum Optics and Quantum Information supplied our entangled photons. About three-quarters of a mile to the north, Thomas Scheidl and his colleagues set up two telescopes in a different university building. One was aimed at the institute, ready to receive the entangled photons, and one was pointed in the opposite direction, fixed on a star in the night sky. Several blocks south of the institute, at the National Bank of Austria, a second team, led by Johannes Handsteiner, had a comparable setup. Their second telescope, the one that wasn't looking at the institute, was turned to the south.

Our group's goal was to measure pairs of entangled particles while ensuring that the type of measurement we performed on one particle of the pair had nothing to do with how we assessed the other. Handsteiner's target star was six hundred light-years from Earth, which meant that the light received by his telescope had been traveling for six hundred years. We selected the star carefully, such that the light it emitted at a particular moment all those centuries ago would reach Handsteiner's telescope first, before it could cover the extra distance to either Zeilinger's lab or Scheidl's station at the university. (In this way, Ellie's waiter would offer her a dessert chosen on the basis of an event that had occurred quadrillions of miles from Earth, which neither Ellie nor Toby nor Toby's waiter could have foreseen.) Scheidl's target star, meanwhile, was nearly two thousand light-years away. Both teams' telescopes were equipped with

Figure 4.2. Johannes Handsteiner sets up equipment on the top floor of the National Bank of Austria to get ready for our first Cosmic Bell test, April 2016. The telescope pointing out the window would gather light from a bright star within our Milky Way galaxy, while equipment on the other side of the hallway would detect and measure entangled photons fired through the night sky from the roof of Anton Zeilinger's laboratory. (*Source*: Photograph by Sören Wengerovsky.)

special filters that could distinguish extremely rapidly between photons that were redder or bluer than a particular reference wavelength. If Handsteiner's starlight in a given instant happened to be more red, then the instruments at his station would perform one type of measurement on the entangled photon, which was just then zipping through the night sky, en route from Zeilinger's laboratory. If Handsteiner's starlight happened instead to be bluer, then the other type of measurement would be performed. The same went for Scheidl's station. The detector settings on each side changed every few millionths of a second, on the basis of new observations of the stars.[13]

By placing Handsteiner's and Scheidl's stations relatively far apart, we were able to close the locality loophole even as we addressed the freedom-of-choice loophole. (Since we detected only a small fraction of all the entangled particles that were emitted from Zeilinger's lab, though, we had to assume that the photons we did measure represented a fair sample of the whole collection.) We conducted two experiments that night, aiming the stellar telescopes at one pair of stars for three minutes, then another pair for three more. In each case, we detected about a hundred thousand pairs of entangled photons. The results from each experiment showed beautiful agreement with the predictions from quantum theory, with correlations far exceeding what Bell's inequality would allow.[14]

How might a devotee of Einstein's ideas respond? Perhaps our assumption of fair sampling was wrong, or perhaps some strange, unknown mechanism really did exploit the freedom-of-choice loophole, in effect alerting one receiving station of what question was about to be posed at the other. We can't rule out such a bizarre scenario, but we can strongly constrain it. To account for our experimental results with some explanation *other* than quantum mechanics, any hypothetical mechanism that could have coordinated all those measurement settings and outcomes would need to have been set in motion before the starlight that our teams observed that night had been emitted. By selecting the measurements to be performed at each receiving station on the basis of events that had occurred long ago and far away, our experiment in Vienna improved upon previous efforts to address the freedom-of-choice loophole by sixteen orders of magnitude, a factor of ten million billion. When the starlight that Handsteiner's group observed that night had first been emitted—six hundred years ago—Joan of Arc was so young, her friends still called her Joanie.

:::

Six hundred years is a long time in human terms, but a mere blink in cosmic history. After all, our observable universe has been expanding for nearly fourteen billion years. On the strength of our result with the Vienna test, Anton was able to secure precious telescope time for our group at the world-class Roque de los Muchachos Observatory on La Palma, in the Canary Islands. The observatory is home to some of the largest optical telescopes in the world. Whereas we had used inexpensive, hobby telescopes for our pilot test in Vienna— Andy, Jason, and I had literally pointed to a back-page ad in *Sky and Telescope* magazine during one of Anton's visits to MIT, to describe the type of equipment we would need for our first test—on La Palma we were able to use enormous telescopes, each with a mirror that stretched four meters across. With such huge light-gathering surfaces, these telescopes could collect light from much fainter, more distant objects.

Our chance came in January 2018. The large telescopes on La Palma are in such high demand among professional astronomers—and our group would need to commandeer two of them, to be used in concert during the same observing periods—that our allotted windows came during the relative off-season for most astronomers. We quickly found out why: our first few scheduled observing nights were washed out, foiled by freezing rain and hail. On our first evening, in fact, the professional telescope operators warned us that

Figure 4.3. Two of the large telescopes at the Roque de los Muchachos Observatory on La Palma. On the left is the Telescopio Nazionale Galileo, which our group used during our Cosmic Bell test in January 2018. (*Source*: Photograph by Calvin Leung.)

if we did not leave the mountaintop right away and retreat to the observatory headquarters (at a modestly lower elevation), then the makeshift road back to headquarters would become impassable for our rental cars, which did not have ice-gripping chains on the tires. Nonetheless, on our final evening at the observatory, we caught a lucky break with the weather. Both telescopes functioned flawlessly, enabling our team to perform real-time measurements of light from two different quasars: monstrous, black-hole-powered primordial galaxies that are so far away that the light we observed that evening had been emitted eight and twelve billion years ago, respectively.

Figure 4.4. Members of the Cosmic Bell collaboration discuss observing options in the control room of the William Herschel Telescope at the observatory on La Palma, January 2018. Anton Zeilinger sits with his back to the camera. Others shown are (*from left to right*) Christopher Benn (leaning), Thomas Scheidl, Armin Hochrainer, and Dominik Rauch. (*Source*: Photograph by the author.)

As in our Vienna test, we generated pairs of entangled photons with a laser, housed in a makeshift laboratory on the mountaintop, and beamed the particles half a kilometer in each direction toward the enormous telescopes.[15] While the entangled particles were in flight, fast electronics at each receiving site took in light from the extragalactic quasars and, on the basis of the color of the quasar observed, prepared to perform one or another measurement on its member of the entangled pair. Once again we found the "spooky" correlations, just as quantum theory predicts. But this time, any alternative mechanism that might have exploited the freedom-of-choice loophole to set up the cor-relations that we measured would need to have been set in motion at least eight billion years ago—long before there was a planet Earth, let alone quantum physicists to ponder the ultimate laws of nature.[16]

:::

Experiments like ours harness some of the largest scales in nature to test its tiniest, and most fundamental, phe-nomena. Beyond that, our explorations could help shore up the security of next-generation devices, such as quantum-encryption schemes, which depend on entanglement to protect against hackers and eavesdroppers.

For me, the biggest motivation remains exploring the strange mysteries of quantum theory. The world described by quantum mechanics is fundamentally, stubbornly dif-ferent from the worlds of Newtonian physics or Einstein-ian relativity. Indeed, quantum theory seems to force us to grapple with questions of chance, randomness, and contin-gency—rather like our study of history. Do events strewn across space and time unfold according to some grand, hid-den plan? Or are we always chasing accidents, beset by un-certainty? If Ellie's and Toby's dessert orders are going to keep lining up so spookily, I want to know why.

CALCULATING

5

From Blackboards to Bombs

Early in the morning of 6 August 1945, a mushroom cloud billowed skyward, hovering eerily above the smoldering city of Hiroshima. Three days later, a similar cloud lingered above Nagasaki. For the first time—and, to date, the only time—nuclear weapons had been used in combat. Days after the bombs were dropped, Japan surrendered and the Second World War lumbered to a close.

The war marked an unprecedented mobilization of scientists and engineers and a turning point in the relationship between science, technology, and the state. By the end of the war, the Allied nuclear weapons project, code-named the Manhattan Engineer District or "Manhattan Project," had enrolled 125,000 people, working at thirty-one secret installations scattered across the United States and Canada. Isotope separation plants in Oak Ridge, Tennessee, stretched the length of a city block; the nuclear reactor facilities at Hanford, Washington, required more than half a billion cubic meters of concrete. Allied efforts on radar, likewise top secret at the time, swelled to comparable scale during the war.[1]

The drama with which the war ended—the detonation of nuclear weapons over cities—cemented the reputation of the Second World War as "the physicists' war." In 1949, for example, *Life* magazine profiled physicist J. Robert Oppenheimer, who had served as scientific director of the wartime Los Alamos laboratory, a central node of the Manhattan Project. Referring to massive military projects like the bomb and radar, the reporter invoked "the popular notion" that the Second World War had been "a physicists' war."[2] By that time the First World War, with its notorious battlefield use of poison gases like chlorine and phosgene, had long since been known as "the chemists' war." The bomb and radar presented a logical counterpoint.

News of the bombings thrust American physicists into the spotlight. As early as May 1946, a commentator in *Harper's* observed, "Physical scientists are in vogue these days. No dinner party is a success without at least one physicist." Scholars who "in the pre-atomic age were considered hopelessly out of touch," the commentator continued, "now find themselves consulted as oracles on questions ranging from the supply of nylons to international organization." Another observer noted soon after the war that physicists young and old—including those who had played no role in the secret, wartime projects—found themselves "besieged with requests to speak before women's clubs" and "exhibited as lions at Washington tea-parties."[3]

Physicists' mundane travels suddenly became draped with strange new fanfare. Police motorcades escorted twenty young physicists on their way to a private conference on Shelter Island, off the northern tip of Long Island, in June 1947; a local booster sponsored a steak dinner en route for the startled guests of honor. B-25 bombers began to shuttle elite physicists-turned-government-advisers between Cambridge, Massachusetts, and Washington, DC, when civilian modes of transportation proved inconvenient. Physics department chairs across the country received a steady inflow of mail throughout the 1950s from grade-school students who now dreamed of becoming nuclear physicists. Other letters, some full of questions and others loaded with theories all their own, streamed in from architects, industrial engineers, Navy officers at sea, prisoners, and patients in tuberculosis wards. By the early 1960s, Americans ranked "nuclear physicist" the third most prestigious profession in a nationwide poll, behind only Supreme Court justice and physician.[4]

Each of these developments seemed to be an inevitable corollary of "the physicists' war." Yet the term originally had nothing to do with bombs or radar and had been introduced as early as November 1941 — weeks before the surprise attack on Pearl Harbor and years before the bombings of Hiroshima and Nagasaki. James B. Conant had explained in a newsletter of the American Chemical Society that the conflict then raging in Europe was "a physicist's war rather than a chemist's."[5] Conant was well placed to know: he was president of Harvard University, chair of the US National Defense Research Committee, and a veteran of earlier chemical weapons projects.[6]

When Conant first wrote about "the physicist's war," no one could know whether the bomb or radar would play any significant role in the war. The Radiation Laboratory, or "Rad Lab," at the Massachusetts Institute of Technology — which served as headquarters for the Allied effort to improve radar — was just one year old. A prototype radar device had recently been rejected by a US Army review board, and funding for the project from the National Defense Research Committee had nearly been revoked. Meanwhile, the Manhattan Project didn't exist yet; Los Alamos still housed a private boys' school. The Army Corps of Engineers requisitioned the site's mud-caked ranch houses to begin setting up the new laboratory several months after Conant had written about "the physicist's war."

Beyond the question of timing, there is the matter of secrecy. Conant oversaw both radar development and the nascent nuclear weapons program; information about each was strictly classified. An experienced, high-ranking government adviser like Conant surely did not intend to disclose some of the nation's most closely guarded secrets in a public newsletter. And there is the nature of the radar and bomb projects themselves. Though each was directed by physicists, they teemed with specialists of many stripes. By the end of the war, physicists were a small minority—only about 20 percent—of the Rad Lab staff. At Los Alamos, the wartime organization chart displayed the various groups (metallurgy, chemistry, ballistics, ordnance, and electrical engineering in addition to physics) arranged in a circle, connected by spindly links. No group appeared on top directing the others. Researchers at both the Rad Lab and Los Alamos forged new kinds of hybrid, interdisciplinary spaces during the war. Neither facility could be categorized simply as a physics laboratory.[7] So what was Conant talking about?

:::

To most scientists and policymakers in the early 1940s, "the physicists' war" referred to a massive, urgent educational mission: to teach elementary physics to as many enlisted men as possible. In January 1942, the director of the American Institute of Physics (AIP), Henry Barton, citing Conant, began issuing bulletins entitled "A Physicist's War." Barton reasoned that "the conditions under which physicists can render services to their country are changing so rapidly" that academic leaders and heads of laboratories needed some means of keeping abreast of evolving policies and priorities.[8] The bimonthly bulletins focused on two main topics: how to secure draft deferments for physics students and personnel and how academic departments could meet the sudden demand for more physics instruction.

Modern warfare, it seemed, demanded rudimentary knowledge of optics and acoustics, radio and circuits. Before the war, the US Army and Navy had trained technical specialists from within their own ranks, at their own facilities. The sudden entry of the United States into the war required new tactics. University physicists, consulting with Army and Navy officials, reported early in the conflict that enrollments in high school physics classes would need to jump 250 percent to meet the new demand. Their goal: half of all high school boys in the country should spend at least one class period per day focusing on electricity, circuits, and radio. The challenge was significant, since at the time less than half of the nation's high schools offered any instruction in physics at all. "New courses in biology and chemistry are not needed," concluded scientists advising officials in the US Office of Education. But the need for physics instruction was urgent: "It is now, each day, each month, this fall, this year," that educators across the country needed to "encourage schools to offer and capable students to take physics," demanded an Office of Education official.[9]

The pedagogical pressures extended beyond high schools. The armed forces also called for massive numbers of military personnel to receive basic physics training at colleges and universities. Draft syllabi circulated between military officials and the AIP. The Army, for example, wanted the new courses to emphasize how to measure lengths and angles, air temperature, barometric pressure and relative humidity, and electric current and voltage. Lessons in geometrical optics would emphasize applications to battlefield scopes; lessons in acoustics would drop applications to music in favor of depth sounding and sound ranging. So acute was the need to teach elementary physics that a special committee recommended in October 1942 that university departments discontinue courses in atomic and nuclear physics for the duration of the war—topics later associated with exotic weapons like nuclear bombs—so as to devote more teaching resources to truly "essential" material.[10]

The material might have been rudimentary, but the pace was grueling. Many colleges and universities shifted from semesters to trimester or quarter systems to fit more courses into a given calendar year. Small liberal arts colleges, such as Williams College in western Massachusetts, began to accept two hundred Navy cadets *per month*; enrollments in the college's physics courses quadrupled. At larger institutions, such as MIT, students in both the Army and Navy programs quickly outnumbered civilian students; by winter 1943–44, the campus hosted three military students for every two civilian ones. Entire buildings on campuses throughout the United States were given over to the special Army and Navy courses and to housing the recruits. Between December 1942 and August 1945, by packing students into two ninety-minute lectures and one three-hour laboratory session each day—six days a week—accelerated courses across the country managed to train a quarter of a million students in elementary physics as part of the Army and Navy programs.[11]

Figure 5.1. Students in the US Army Special Training Program attend lectures at MIT, ca. 1944. (*Source*: MIT *Technique* magazine [1944], courtesy of MIT *Technique* editorial board.)

Staffing the inflated classrooms required military-style planning and logistics. Barton's AIP bulletins warned that any universities found to be hoarding valuable physics teachers—much less poaching them from other schools—would be subject to "severe criticism." Barton developed a complicated formula for what he termed the acceptable ratio of "genuine to 'ersatz' teachers of physics" in any given institution.[12] Departments deemed to have higher ratios of "genuine" (experienced) physics teachers than Barton's formula allowed would be subject to censure. Specialists in neighboring fields, such as mathematics, chemistry, and engineering, were to be "drafted" to begin teaching physics instead. At small colleges like Williams, an even wider mix of faculty pitched in. After some quick retooling, professors from music, theater, philosophy, geology, and biology helped to teach physics to the Navy recruits.[13] Physics teachers became a rationed commodity: like rubber, gasoline, and sugar, they were in critically short supply.

With undergraduate physics enrollments slated to triple by the end of 1943, draft policies quickly followed suit. The US government created the National Committee on Physicists in December 1942—the first of its kind for any academic specialty—to advise local draft boards on the need for teaching-related deferments. Soon the phrase "the physicists' war" echoed throughout newspapers, popular magazines, and even congressional testimony. Use of the phrase peaked in 1943, long before there was much news to report (classified or otherwise) about the Manhattan Project.

:::

After the war, use of the phrase "physicists' war" rebounded every decade or so, usually around an anniversary of the bombings of Hiroshima and Nagasaki. The highest postwar peak accompanied the publication of Richard Rhodes's Pulitzer Prize–winning book *The Making of the Atomic Bomb* in 1986.[14] By then, Conant's phrase had long since been linked with classified military projects rather than classroom instruction.

The transition began nearly as soon as the bombs were

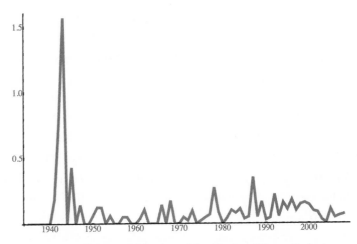

Figure 5.2. Google *n*-gram of the phrase "physicists' war." The vertical axis shows percentage (× 10⁶) of occurrences of the bigram "physicists' war" among all bigrams in the available English-language corpus. (*Source*: Figure by the author, based on data at https://books.google.com/ngrams.)

dropped over Japan. Soon after work on the Manhattan Project had begun, its main overseer, General Leslie Groves, anticipated that the government would eventually need to have some information ready to release about the top-secret nuclear weapons project—pre-cleared and available for wide distribution—in case the bombs were ever used. He tapped Princeton nuclear physicist Henry DeWolf Smyth to spend the war visiting each Manhattan Project site, compiling a technical report on the project that would be suitable for public dissemination.[15]

On the evening of 11 August 1945, just two days after the bombing of Nagasaki, the US government released Smyth's two-hundred-page document under the ponderous title *A General Account of Methods of Using Atomic Energy for Military Purposes under the Auspices of the United States Government, 1940–1945.* Quickly dubbed the "Smyth report," copies flew off the shelves. The original Government Printing Office edition ran out so quickly that Princeton University Press soon published its own edition, under the more manageable title *Atomic Energy for Military Purposes*.[16]

As historian Rebecca Press Schwartz has documented, security considerations dictated what Smyth could and could not include in his report. Only information that was already widely known to working scientists and engineers or that had "no real bearing on the production of atomic bombs" was deemed fit for release. Little of the messy combination of chemistry, metallurgy, engineering, or industrial-scale manufacturing met these criteria; those aspects of the huge project that were crucial to the actual design and production of nuclear weapons remained closely guarded.[17] Instead, Smyth focused narrowly on ideas from physics, pushing theoretical physics in particular to the forefront. Ironically, most people read in Smyth's report the lesson that physicists had built the bomb—and, by implication, had won the war.[18]

It needn't have been that way. Consider, for example, a long-forgotten press release issued by the War Department on 6 August 1945, the day that Hiroshima was bombed. This particular release was issued locally, in the state of Washington, and it played to the home crowd. More than two tightly packed pages extolled the great breakthroughs in chemistry and chemical engineering that had made the bomb possible—breakthroughs specifically associated with the massive Hanford site in Washington. Not one word of this particular press release referred to physics or physicists.[19]

Yet rare examples like this press release could hardly compete with the overwhelming attention devoted to the Smyth report. Soon after Princeton University Press brought out its edition of the report, Smyth's book spent fourteen weeks on the *New York Times* best-seller list; it sold more than a hundred thousand copies in a little over a year.[20] Later reports, such as *Essential Information on Atomic Energy*, issued in 1946 by the new Special Committee on Atomic Energy of the US Senate, borrowed liberally from the Smyth report, depicting nuclear weapons as the latest in a series of developments in theoretical physics. A "chronological table" at the end extended the narrative as far back as 400 BC to the ancient Greek atomists—rather than, say, to the Berlin chemistry laboratory in which nuclear fission had been discovered late in 1938, much less to the work of DuPont chemical engineers who had managed to scale up plutonium-producing nuclear reactors during the war.[21]

:::

The change in referent for the phrase "the physicists' war"—from blackboards to bombs—had serious implications. After the war, physicists in the United States bore the largest brunt of any academic group during the McCarthy-era red scare. The House Un-American Activities Committee held twenty-seven separate hearings on allegations against physicists, twice the number for members of any other scholarly discipline. If nuclear weapons had been made by physicists, so the reasoning went, then physicists must have special access to the "atomic secrets" with which such bombs could be made. Hence the loyalties of this group required the closest scrutiny of all.[22]

On the other hand, many policymakers quickly concluded after the war that if physicists really did hold the atomic secrets, then the United States needed many more physicists to secure the uneasy peace. Just six weeks after the war had ended, the Scientific Panel of the Interim Committee on Atomic Power—the selfsame group that had advised the secretary of war on possible uses for the new weapons during the war—recommended to General Groves that the Army continue to support university physics research, a plan that Groves quickly approved. Groves and his colleagues were eager "to prevent disintegration of [the Army's] nuclear research organization" and hence to bolster "advanced training in nuclear studies."[23] The Office of Naval Research drew similar conclusions, investing heavily after the war in unclassified research projects so as to make up for the "deficit in technical people" left in the wake of the recent conflict. An official from the Office of Naval Research explained during a meeting with advisers at the Pentagon in 1947 that "graduate students working part time are slave labor"—hence, supporting the training of more physics graduate students was an easy investment for the Navy to justify.[24]

The following year, as Congress oversaw a new fellowship program to encourage students to pursue graduate study of physics, the program's director concurred with a leading senator that the program should "establish a nucleus of highly trained individuals who will increase the general knowledge in scientific fields and at the same time provide a pool from which some individuals will be drawn to active employment on the atomic energy program." The plan worked: by 1953, three-quarters of the graduate students who completed their PhDs in physics with funding from the Atomic Energy Commission—the postwar successor to the Manhattan Project, which assumed control of the nation's nuclear arsenal and oversaw research and development for the sprawling nuclear complex—took jobs with the commission upon graduation.[25]

These postwar calculations had an immediate impact on how research in areas like physics was supported. In 1949,

96 percent of all funding for basic research in the physical sciences within the United States came from defense-related federal agencies, including the Department of Defense and the Atomic Energy Commission. In 1954—four years after the establishment of the civilian National Science Foundation—fully 98 percent of funding for basic research in the physical sciences came from federal defense agencies. The scale of funding, much like the source, bore little resemblance to the prewar patterns. By 1953, funding for basic research in physics in the United States was twenty-five times higher than its level in 1938 (in constant dollars).[26]

With these vast infusions of spending, enrollments in physics skyrocketed. At a time when higher education was booming across all fields of study—the G.I. Bill brought more than two million veterans into the nation's colleges and universities after the war—graduate enrollments in physics grew at the fastest pace of all, doubling almost twice as quickly as the rate for all fields combined. By the outbreak of fighting in the Korean War in June 1950, American physics departments were producing three times as many PhDs in a given year as the prewar highs.[27] Weeks into the new conflict, officials with the National Research Council and the AIP scrambled to make sure those trends would continue. They hastily drew up a memorandum to establish "procedures for utilizing our manpower in physics," emphasizing, above all, that "unless a very short emergency (not more than three years) is contemplated, it is of great importance to continue the training of promising students to full competence in physics." After all, the administrators warned, "we have now no 'stockpile' of physicists." Many heeded the new call. Citing the nation's entry into the Korean War, the physics department at the University of Rochester immediately began offering four additional teaching assistantships and five new research assistantships to encourage more students to pursue graduate training in physics.[28]

Just as they had during the Second World War, scientists and policymakers tracked the postwar training effort carefully, fearful that despite the rapid rise in physics enrollments the nation might suffer some critical shortfall. In the course of a single speech in 1951, Smyth—now in his role as a top member of the Atomic Energy Commission—described young physics graduate students as a "war commodity," a "tool of war," and a "major war asset," to be "stockpiled" and "rationed." Analysts at the Bureau of Labor Statistics agreed. "If the research in physics which is vital to the nation's survival is to continue and grow," they asserted in a 1952 report, "national policy must be concerned not only with keeping young men already in the field at work but also with insuring a continuing supply of new graduates." Similar calculations produced huge numbers of newly trained physicists in other Cold War powers as well, including the United Kingdom and the Soviet Union. In fact, more physicists were trained during the quarter century after the Second World War than had ever been trained, cumulatively, in human history.[29]

The language of "rationing" and "stockpiling" young physicists thus percolated with little interruption, starting in the closing weeks of 1941—before the Manhattan Project even existed—and continuing throughout the first decade of the nuclear age. However, although the terminology remained fairly constant, the aims of the training shifted considerably after the war. Rather than teach soldiers some elementary physics to prepare them for the battlefield, US officials began to talk about creating a "standing army" of physicists, specialized in the esoterica of nuclear reactions, who could work on weapons projects without delay should the Cold War ever turn hot.[30] By the mid-1950s, Conant's phrase about "the physicists' war" had assumed a new meaning and a new urgency. But one underlying point remained unchanged: it was all about training, after all.

6

Boiling Electrons

Two decades ago, while digging through a physicist's archive, I stumbled upon a document that has haunted me
ever since. It was a hand-typed table of integrals—long lists
of mathematical functions and the numerical answers one
should find upon integrating the functions between various limits—seemingly little different from the manuals and
tables that I had kept handy as a student when trying to
solve my homework problems. The familiarity of the contents jarred with the table's front page. Precisely thirty-one
copies of the table had been printed, their recipients carefully noted on the cover. The table, dated 24 June 1947, had
been prepared to accompany a classified report. The distribution lists for the two documents were a close match;
nearly all the recipients of the integral table (like all who received the main report) had attained security clearances to
handle secret, defense-related materials.[1]

How to reconcile the banal contents with the striking

cover page? What disaster would have befallen the US government if enemies of the state had learned that the integral of $x/(1 + x)^2$ between $x = 0$ and $x = 1$ equaled 0.1931? Moreover, how could the authorities have hoped to thwart the circulation of such basic mathematical results? Wouldn't anyone schooled in the routines of calculus arrive at the same answers, whether or not their names appeared on the table's special distribution list?

The table of integrals was prepared as a supplement to a classified report written by renowned physicist and Nobel laureate Hans Bethe. (I found both items in Bethe's papers at Cornell University in upstate New York.) In the 1930s, Bethe had become one of the world's experts on nuclear physics; by 1938, he had pieced together the complicated nuclear reactions that make stars shine. He served as the director of the Theoretical Physics Division at wartime Los Alamos, reporting directly to J. Robert Oppenheimer. After the war, when he returned to teaching at Cornell, he remained an active consultant to the nuclear weapons program as well as to the budding nuclear power industry.[2]

In 1947, Bethe had been asked to work on the problem of shielding for nuclear reactors. When heavy nuclei like uranium or plutonium get blasted apart by neutrons, they release energy—energy that can power a bomb or generate electricity in a reactor—and they also release large doses of high-energy radiation. While working on the challenge of how best to block or absorb the radiation, Bethe kept finding that he needed to evaluate integrals of a particular form. A colleague—another PhD physicist, Manhattan Project veteran, and by then senior researcher at a nuclear reactor facility—prepared the accompanying table of integrals so that selected coworkers might be able to perform calculations like Bethe's.

Similar mathematical handbooks and tables had been prepared for centuries. Around the time of the French Revolution, for example, as historian Lorraine Daston has written, leading civil servants produced mammoth tables of logarithms and trigonometric functions calculated to fourteen or more decimal places—far greater accuracy than any practical application would have required at the time. Gaspard Riche de Prony's tables were a deliberate demonstration of Enlightenment mastery, one more testament to the triumph of Reason, to be admired more than used.[3]

Though the 1947 table of integrals was not prepared for public fanfare—just the opposite, as its closely watched distribution list made clear—the table stood closer to Prony's time than to ours. Indeed, the table's introduction explained the provenance of many of the results that would follow: most integrals had been evaluated by making clever changes of variables so that the functions of interest matched forms that had been reported in the venerable *Nouvelles tables d'intégrales définies*, published in Leiden in 1867 by the wealthy Dutch mathematician David Bierens de Haan.[4]

Two years into the atomic age, the labor of calculation still resembled the virtuosity of the humanist scholar more than the pragmatics of the engineer. One needed access to a well-stocked library filled with old, foreign-language books. And here lies a key to understanding the limited distribution of the 1947 integral table: although in principle anyone should have been able to compute the integrals, completing such calculations in practice required substantial resources of time and skill. Soon after typists finished preparing Bethe's report and the accompanying integral table, however, the nature of calculation would change forever.

The unlikely setting for the transformation was the Institute for Advanced Study in Princeton, New Jersey. The

Institute had been founded in 1930 with generous funding from department-store magnate Louis Bamberger and his sister, Caroline Bamberger Fuld. Advised by education reformer Abraham Flexner, the founders aimed to establish a place for budding young intellectuals to pursue their scholarship beyond their doctorates, before the routines of university life—teaching, committee work, and all that—could dampen their creativity. Flexner sought some quiet place where scholars would have no institutional obligations to write lectures or publish reports; a place where they could sit and think. "Well, I can see how you could tell whether they were sitting," quipped leading science policymaker Vannevar Bush.[5]

Flexner began to staff the Institute with world-famous intellectuals who were fleeing fascism in Europe. Albert Einstein became the first permanent faculty member in 1933; he was soon joined by eccentric, lonely geniuses like logician Kurt Gödel, who eventually starved himself to death out of paranoid fear that people were trying to poison his food. Even after Oppenheimer became director of the Institute in 1947, fresh from his frenzied work as director of Los Alamos, the Institute remained closer in spirit to a monastery than a laboratory—a place far more likely to stack leather-bound copies of Bierens de Haan's old *Nouvelles tables d'intégrales* on its shelves than to host the whir of lathes and drills. A *New Yorker* reporter observed in 1949 that the Institute had a "little-vine-covered-cottage atmosphere." Around that time Hans Bethe advised a young physicist who was about to relocate to the Institute for a year "not to expect to find too much going on" there.[6]

The calm began to be disturbed by members of a new team assembled by legendary mathematician John von Neumann. Born in Budapest in 1903, von Neumann published his first mathematical research papers at the tender age of nineteen and completed his doctorate at twenty-two, before fleeing the Continent as Europe descended into turmoil. Flexner scooped him up and added him to the Institute's roster soon after Einstein had joined. Von Neumann spent much of the war at Los Alamos, working alongside Bethe and Oppenheimer on nuclear weapons. While immersed in that work, he became gripped by a vision as remarkable as Charles Babbage's a century before: perhaps one could build a machine to calculate. Von Neumann was motivated not only by curiosity about the working of the human brain and the essence of cognition (though he was fascinated by such topics). He had more pressing concerns as well: he needed to know whether various designs for nukes would go boom or bust.[7]

The wartime weapons project gave von Neumann a taste for semiautomated computation. Among the challenges he and colleagues faced was tracking, in some quantitative way, the likely outcomes when neutrons were introduced into a mass of fissionable material. Would they scatter off the heavy nuclei, get absorbed, or split them apart? Equally important: how would a shock wave from exploding charges propagate through the bomb's core? During the war, calculations like these were largely carried out by chains of human operators armed with handheld Marchant calculators, a process recounted in David Alan Grier's fascinating book *When Computers Were Human* (2005). Young physicists like Richard Feynman carved up the calculations into discrete steps, and then assistants—often the young wives of the laboratory's technical staff—would crunch the numbers, each assistant performing the same mathematical operation over and over again. One person would square any number given to her; another would add two numbers and pass the result to the next woman down the line.[8]

That rough-and-ready process had worked well enough for wartime calculations pertaining to fission bombs. But hydrogen bombs were a whole different beast—not just in potential explosive power but computationally as well. Their internal dynamics, driven by a subtle interplay between roiling radiation, hot plasma, and nuclear forces, were far more complicated to decipher. Trying to determine whether a given design would ignite nuclear fusion—forcing light-weight nuclei to fuse together as they do inside stars, un-leashing thousands of times more destructive power than the fission bombs that were dropped on Hiroshima and Nagasaki—or whether it would fizzle posed tremendous computational challenges. Such calculations could never be completed by teams of people brandishing Marchant calcu-lators. They required, or so von Neumann concluded, a fully automated means of solving many complicated equations at once. They required an electronic, digital computer that could execute stored programs.[9]

Some of the original ideas behind stored-program com-putation had been invented before the war by British mathe-matician and cryptologist Alan Turing, and indeed, the world's first instantiation of Turing's ideas was completed by a team in Manchester in 1948. Much like the Manhat-tan Project and the wartime radar program, however, what had begun as a British idea was scaled up to industrial size by the Americans. One group, centered at the University of Pennsylvania and sponsored by the US Army, had been hard at work on a similar device since 1943, code-named ENIAC for "Electronic Numerical Integrator and Computer." Im-mediately after the war the Pennsylvania group gained competition, as von Neumann began to amass government contracts to build his own computer at the Institute in Princeton. His team included several young engineers as well as his talented wife, Klári, who dove into the minutiae of coding bomb simulations and coaxed the machine along as it ran for days on end.[10]

Von Neumann had rubbed shoulders with Turing at the Institute during the 1930s, while Turing worked on his dissertation at nearby Princeton University. During the war, von Neumann had also consulted on the ENIAC in Pennsylvania. In fact, he helped to redirect the project from its original mandate—calculating artillery tables for the Army's ballistics laboratory—to undertaking calculations on behalf of Los Alamos for its nuclear weapons designs. At the time, the Pennsylvania machine could execute only fixed programs: one had to set a given program by hand by physically rewiring components ahead of time, before any results could be calculated. Changing the program took weeks of physical manipulation—swapping cables, alternating switches, checking and rechecking the resulting combinations. Like the ENIAC's inventors, von Neumann sought some means by which a computer could store its program alongside the resulting data, in the same memory. Just as Turing had envisioned, such a machine would store its instructions and its results side by side.

Designed before the invention of the transistor, von Neumann's computer required more than two thousand vacuum tubes to work in concert. Such tubes were already a decades-old technology. They produced electric current by boiling electrons off a heated chunk of metal—hardly the image of today's sleek laptops or smartphones. To counter the constant heating from the tubes, the machine required a massive refrigeration unit, capable of producing fifteen

Figure 6.1. John von Neumann (*left*) talks with J. Robert Oppenheimer, director of the Institute for Advanced Study, near a portion of the Institute's electronic computer, ca. 1952. (*Source*: Photograph by Alan Richards, courtesy of the Shelby White and Leon Levy Archives Center, Institute for Advanced Study, Princeton, NJ, USA.)

tons of ice per day. With astounding dexterity, von Neumann's small but energetic team brought their room-sized computer to life during the late 1940s. By the summer of 1951, the machine was chugging full-time on H-bomb calculations, running around the clock for two-month stretches at a time. When operating at full capacity, the Institute computer could boast 5 kilobytes of usable memory. As George Dyson notes in his engaging account of von Neumann's project, that's about the same amount of memory that today's computers use in half a second when compressing music files.[11]

The Institute computer project was fueled largely by contracts from the Atomic Energy Commission, the postwar successor to the Manhattan Project. Those contracts stipulated that virtually no information about thermonuclear reactions could be released to the public—a position that President Harry S. Truman reiterated when, on the last day of January 1950, he committed the United States to crash-course development of an H-bomb. As a cover for their main task, therefore, von Neumann's team also worked on unclassified problems as they put their new machine through its paces. Weather prediction became a popular challenge. Meteorology featured many of the kinds of complicated fluid flows that weaponeers also had to understand inside their H-bomb designs. Other early applications of the computer included simulations of biological evolution, whose processes branched like so many scattered neutrons inside a nuclear device.

At the end of the 1950s, physicist and novelist C. P. Snow diagnosed a clash between "two cultures": literary intellectuals versus natural scientists.[12] At the Institute, von Neumann's electronic monster induced a sharp clash of cultures, but not along the fault lines that Snow had in mind. Rather, the gap yawned between notions of an independent scholarly life and the teamwork regimen of engineers. By 1950, the budget for von Neumann's computer project, drawn almost entirely from government defense contracts, dwarfed the entire budget for the Institute's School of Mathematics. More than mere money was at stake; so, too, were ways of life. As Institute mathematician Marsten Morse wrote to a colleague in the early 1940s, "In spirit we mathematicians at the Institute would cast our lot in with the humanists." Mathematicians, Morse continued, "are the freest and most fiercely individualistic of artists."[13] Morse's Institute neighbor, Einstein, agreed. Reviewing an application from a young physicist to the Guggenheim Foundation in 1954, Einstein conceded that the proposed topics of study were worthy, but he considered the overseas trip unnecessary: "After all everybody has to do his thinking alone."[14] At the Institute, the computer project did not pit science against the humanities; the battle was between ideals of the Romantic Genius versus the Organization Man.

That difference in temperament was expressed in more tangible ways as well. The computer project was originally housed in the basement of the Institute's main building, Fuld Hall—out of sight, even if the clanging did not keep the engineers quite out of mind. Soon the computer group was moved to facilities further from the Institute's more solitary scholars. But even the new building required a telling compromise. The drab, functional, concrete building envisioned by the government sponsors would never have satisfied the aesthetes among the Institute's main residents, so the Institute paid an additional $9,000 (nearly $100,000 in today's currency) to cloak the new building with a brick veneer.

In the end, the Institute's computer project became a victim of its success. Von Neumann moved more squarely into policymaking for the atomic age when, in 1955, he was tapped to serve as one of the five members of the Atomic Energy Commission—soon after the commission had infamously stripped von Neumann's boss, Oppenheimer, of his security clearance. (Much of Oppenheimer's hearing had turned on his counsel against H-bomb development.) Oppenheimer remained in charge of the Institute, even as von Neumann spent less and less time there. With von Neumann away from campus, the computer project had no local lion to protect its turf. Other centers across the country, meanwhile, began to make fast progress on replica machines. One of them, at the US Air Force–funded think tank RAND, was even dubbed the "Johnniac," in honor of von Neumann. Von Neumann succumbed to cancer in 1957; the Institute computer project limped along for a few more months until it was finally shuttered in 1958. By that time, computation no longer relied upon solitary scholars poring over dusty reference volumes. The computer age had arrived.[15]

I reread Bethe's 1947 memo and the accompanying table of integrals a few summers ago while traveling in rural Montana. A tire on my rental car had picked up a nail somewhere between Bozeman and Kalispell. While I waited for a mechanic to get me roadworthy again, I flipped open my laptop and began to reevaluate some of the integrals that had been so carefully guarded sixty-five years earlier. These days run-of-the-mill software on typical laptops can evaluate such integrals in microseconds—the limiting factor is one's typing speed rather than the processing power of the machine. Whereas the 1947 table listed numerical answers to four decimal places, my laptop spit back answers to sixteen decimal places quicker than I could blink. A few more keystrokes and I could have any answer I wanted to thirty decimal places. No need for access to a fancy library; no need to slog through a savant's treatise in Dutch. I could sit in the service station, aside a dusty back road, and calculate.

7

Lies, Damn Lies, and Statistics

"Russia Is Overtaking U.S. in Training of Technicians," blared a typical front-page headline in the *New York Times* in 1954. "Red Technical Graduates Are Double Those in U.S.," echoed the *Washington Post*.[1] Even after news of once-secret wartime efforts like radar and the Manhattan Project had spurred lightning-fast growth in physics enrollments across the country, many feared that the nation's technical workforce remained insufficient to meet the demands of the Cold War. Soviet advances in training young physicists seemed especially menacing. Widespread fears that the Soviet Union had surpassed the United States in training large cadres of physical scientists drove a massive escalation in American efforts to train more specialists. Graduate enrollments in physics surged even more quickly after the surprise launch of the Soviet *Sputnik* satellite in October 1957. Yet the runaway growth proved unsustainable. Less than fifteen years after *Sputnik*, funding, enrollments, and job opportunities for young physicists in the United States collapsed.

Looking back at those wild swings today, a single pattern emerges. Whether one plots research funding, graduate-level enrollments, or job listings in physics over time, the curves all look the same: up races the curve until all at once the bottom falls out, crashing as sharply as the headlong rise had been. To our eyes today, in fact, such curves seem eerily familiar. Change the labels and the same graph could just as well describe a tech-stocks boom or a housing-market bust. During the Cold War, in other words, physics training in the United States had become a speculative bubble.

Economist Robert Shiller defines a speculative bubble as "a situation in which temporarily high prices are sustained largely by investors' enthusiasm rather than by consistent estimation of real value." He emphasizes three distinct drivers for bubbles: hype, amplification, and feedback loops. Consumers' enthusiasm for a particular item—be it an initial public offering for the latest Silicon Valley start-up or a hip loft in SoHo—can attract further attention to that item. Increased media attention, in turn, can elicit further consumer investment, and the rise in demand will drive prices up even further. Before long, the price increase can become a self-fulfilling prophecy. "As prices continue to rise, the level of exuberance is enhanced by the price itself," Shiller explains.[2]

As with stock prices, so with graduate training. The skyrocketing enrollments in fields like physics during the Cold War were fed by earnest decisions based on incomplete information, intermixed with hope and hype that had little grounding in fact. Feedback loops among scientists, journalists, and policymakers kept the demand for young physical scientists artificially inflated. Faulty assumptions that could easily have been checked instead came to seem natural, even inevitable, when refracted by geopolitical developments. When those conditions changed abruptly, physics had nowhere to go but down.

::::

My favorite example of the hype-amplification-feedback process concerns a series of reports that were undertaken during the 1950s on Soviet advances in training scientists and engineers. As early as 1952, while the Korean War smoldered, several analysts began trying to assess Soviet "stockpiles" of scientific and technical manpower—those cadres who seemed so "essential for survival in the atomic age," as one *New York Times* reporter put it. Three lengthy reports were prepared to assess what became known as the "cold war of the classrooms": Nicholas DeWitt's *Soviet Professional Manpower* (1955), Alexander Korol's *Soviet Education for Science and Technology* (1957), and DeWitt's *Education and Professional Employment in the USSR* (1961).[3]

The three studies shared many features. Each was conducted in Cambridge, Massachusetts, by researchers who had themselves undergone some of their training in Russia and the Soviet Union. Nicholas DeWitt completed the first report, *Soviet Professional Manpower*, while working at Harvard's Russian Research Center. The center had been established in 1948 with aid from the US Air Force and the Carnegie Corporation; throughout this period it also maintained close ties with the Central Intelligence Agency. DeWitt, a native of Kharkov, in Ukraine, had begun his training at the Kharkov Institute of Aeronautical Engineering in 1939, before the Nazi invasion forced him to flee. He eventually landed in Boston in 1947 and enrolled as an undergraduate at Harvard the following year. In 1952, honors degree in hand, DeWitt began working as a research associate at the Russian Research Center while pursuing graduate study at Harvard in regional studies and economics. The National Science Foundation and the National Research Council jointly sponsored his investigation into Soviet scientific and technical training. Colleagues called him compulsive, an "indefatigable digger," and it showed: his massive follow-up study, *Education and Professional Employment in the USSR*, ran to 856 pages, featuring 257 tables and 37 charts in the main text alone, followed by 260 dense pages of appendices.[4]

The other major report, Alexander Korol's *Soviet Education for Science and Technology*, took shape down the street at MIT's Center for International Studies. Like the Russian Research Center at Harvard, MIT's center (founded in 1951) also maintained close ties to the CIA, which secretly bankrolled Korol's study. Korol, like DeWitt, was a Soviet expatriate who had first trained in engineering. Korol enlisted aid from several MIT faculty in the sciences and engineering to help him gauge the quality of Soviet pedagogical materials. He completed his study in June 1957; the book's preface, by the center's director Max Millikan, was dated 18 October 1957, barely two weeks after the surprise launch of *Sputnik*. The report was immediately heralded as "fastidious," "perhaps the most conclusive study ever made of the Soviet education and training system." Others marveled at the "400 pages of solid factual data" crammed between the book's covers.[5]

Both authors carefully emphasized caveats and qualifications. DeWitt began both of his books by citing the large professional literature devoted to interpreting Soviet statistics. Both of his books also included detailed appendices on the "perplexities and pitfalls" of working with Soviet statistics. Raw data like enrollment figures or graduation rates never speak for themselves, DeWitt cautioned; such social statistics always require careful interpretive work. All the more so in the Soviet case: mundane gaps in data (which afflicted most social-scientific studies) were compounded by the Soviet government's penchant for secrecy and for propagandistic massaging of data. Korol similarly urged caution, arguing time and again that it was fruitless to compare graduation rates between the Soviet Union and the United States, since the two nations' educational systems differed so radically in structure and function. Indeed, Korol refused even to tabulate Soviet and American statistics side by side, in order to avoid "unwarranted implications."[6]

DeWitt and Korol urged that Soviet educational trends be seen in the proper light. Although curricula for elite programs of study—such as physics at Moscow State University and at Columbia University or MIT—seemed to be roughly comparable in quality, several mitigating factors stood out. First, they both argued, a large proportion of scientists and engineers in the Soviet Union never practiced their craft, working instead in various bureaucratic or administrative posts. Second, the Soviet system was built around extraordinary specialization: the specialty of nonferrous metals metallurgy, for example, was itself carved up into eleven distinct specializations (copper and alloys metallurgy, precious metals refining, and so on). Students selected only one narrow specialty and devoted the bulk of their studies to it. Well into the late 1950s, meanwhile, Soviet students suffered from widespread shortages of textbooks and poor-quality (or missing) laboratory equipment. Student-to-faculty ratios had ballooned immediately after the war and continued to widen over the 1950s. There were also indications that academic standards were massaged to fit the central planning committee's "production quotas": both Korol and DeWitt noted internal Soviet reports of pressure to let mediocre students pass when overall numbers looked low.[7]

Most important of all, a fast-growing proportion of Soviet students were enrolled in extension-school or correspondence programs. Unlike regular full-time students, these students held full-time jobs away from universities and pursued their studies largely alone, reading textbooks (when these were available) and occasionally sending written assignments to overworked professors, most of whom juggled sixty-five to eighty such students at a time. Even Soviet education officials routinely remarked on the inferior quality of this type of training, especially for hands-on fields like science and engineering. Yet enrollments in extension and correspondence programs were soaring, even as regular full-time enrollments remained flat. By 1955, extension and correspondence students composed about one-third of all engineering enrollments in the Soviet Union; five years later, they accounted for more than half of enrollments in all fields combined.[8]

Only after delineating each of these factors at length did DeWitt broach numerical comparisons. Focusing on the Soviet five-year "diploma" programs—roughly akin to American-style training at the undergraduate and master's degree levels—he presented some quantitative findings. Total enrollments in the United States were substantially greater than in the Soviet Union: three times as great as the regular full-time student population in 1953–54, for example, and still one-third larger if one included all the extension and correspondence students in the Soviet tally. Yet the balance of fields was quite different. In the United States, roughly one out of four students majored in scientific or technical fields, while in the Soviet Union it was three out of four. In particular, when DeWitt counted up annual degrees granted in the two countries, it appeared that the Soviets were graduating two to three times more students per year in science and engineering than were American institutions.[9]

That ratio—"two to three times"—quickly took on a life of its own. DeWitt's and Korol's reports had been careful, lengthy, serious affairs. The journalistic coverage, on the other hand, leaned toward the sensationalistic. Major newspapers like the *New York Times* and the *Washington Post* splashed the "two to three times" finding across their front pages. Leading spokespeople from the CIA, the Department of Defense, the Joint Congressional Committee on Atomic Energy, and the Atomic Energy Commission trotted out the same stripped-down number in public speeches and congressional testimony, with no trace of DeWitt's caveats or cautions. Each proclamation elicited further hand-wringing in the newspapers.[10] Here, in raw form, was the first step in economist Robert Shiller's model of a speculative bubble: hype.

Before *Sputnik*, at least some observers tried to put the numbers in perspective, much as DeWitt had urged all along. In June 1956, Lee DuBridge—former scientific director of the wartime Radiation Laboratory at MIT, which had served as headquarters of Allied efforts in radar, and at the time president of Caltech—addressed the media frenzy when he testified before the newly formed National Committee for the Development of Scientists and Engineers. (Eisenhower had established the elite twenty-one-member committee just two months earlier, in response to congressional hectoring on the scientific manpower issue.) "It is true that in Russia more men and women received degrees in science and engineering last year than in the United

States," DuBridge began. "So what? Maybe that is because in the past 100 years they have so neglected their technical strength that they must now exert strenuous efforts to build it up. If this is true, then our rate of production should not be determined by their weakness—only by our own. Let us ask how many engineers we need to do our job, and not take over their figures from the numbers they require to do their job." DuBridge might have mentioned another of DeWitt's findings to bolster the point: even after the Soviets' recent burst in scientific and technical training, they still lagged behind the United States in accumulated numbers of scientists and engineers available to the workforce.[11]

More typical, however, was the lesson that Senator Henry "Scoop" Jackson read in DeWitt's numbers. Jackson released a special report entitled "Trained Manpower for Freedom" on 5 September 1957, just one month before *Sputnik* revved the rhetoric of manpower still higher. With the Soviets marching forward on the scientific manpower front—he invoked the now-familiar ratio—Jackson urged that nothing was "more precious" than the immediate development of all potential scientific talent in the United States and its NATO allies. More than ten pages spelled out various training programs to address the critical shortfall, including fellowships for high school and college students, special summer study institutes, and awards for students and teachers who excelled in science education. These resources, Jackson explained with a telling metaphor, "should be used as catalytic agents which, so to speak, can initiate educational chain reactions extending over the broadest possible scientific and technological front."[12]

Sputnik further galvanized these discussions. DeWitt despaired of the "hysterical" reaction sweeping the country; it must have been especially galling to hear his own statistic echoed over and over again, stripped of all nuance and subtlety. Responding to the satellite, for example, former president Herbert Hoover groaned that "the greatest enemy of all mankind, the Communists, are turning out twice or possibly three times as many" scientists and engineers as the United States. Senator Lyndon Johnson quickly convened hearings before his Senate Defense Preparedness subcommittee within a week of the satellite's launch, before which General James Doolittle (famous for his Tokyo bombing raid during the Second World War) brandished the same dire figure. During closed sessions of the hearings, CIA director Allen Dulles returned to the "manpower gap." Details of Dulles's testimony remained secret, but Johnson alerted the press that Dulles had confirmed that the Soviets were "now outstripping the U.S. in developing a scientific and technological manpower pool."[13] In the frenzied weeks after *Sputnik*, Korol's book suffered similar misreadings. Reporting on the book's release, one *Washington Post* article began by exclaiming, "The free world must radically change its ways to meet the challenge of the Soviet Union's power to marshal brains and resources for priority projects." This was an exact inversion of Korol's point—as he had been at pains to make clear—and yet the reporter attributed this alarmist conclusion to Korol himself. Another *Post* article interwove coverage of Korol's book with quotations from Eisenhower's post-*Sputnik* addresses, giving the appearance that both had called for the "absolute necessity of increasing our scientific output" in trained personnel.[14]

Next came Shiller's second phase: amplification. Enterprising physicists took full advantage of the opportunity afforded by the launch of *Sputnik* to further flog DeWitt's number. Quickest to respond was Columbia University's pugnacious Nobel laureate I. I. Rabi. Eisenhower and Rabi had known each other since the late 1940s, when Eisenhower had served as Columbia's president; after Eisenhower became president of the United States, Rabi headed his new Science Advisory Committee. Meeting just a week and a half after the launch of *Sputnik*, Rabi pressed Eisenhower to use the satellite as a pretext for bulking up American scientific training. Soon after that, Elmer Hutchisson, the new director of the American Institute of Physics (AIP), a large umbrella organization that helped to coordinate various professional societies in the field, opined to reporters from *Newsweek* that the entire American way of life could well be "doomed to rapid extinction" unless the nation's scientific reserves were expanded quickly. Behind the scenes, Hutchisson alerted his AIP colleagues that they had "the opportunity of influencing public opinion greatly." He saw "an almost unprecedented opportunity," he wrote in a memo to the AIP Advisory Committee on Education, "to take advantage of the present public questioning concerning the quality of science instruction in our schools." Edward Teller, a veteran of the secret H-bomb program who had helped Senator Jackson prepare his "Trained Manpower" report, hit the same theme when talking with the press. "We have suffered a very serious defeat," he exclaimed, "in a field where at least some of the most important engagements are carried out: in the classroom." Hans Bethe from Cornell University—Los Alamos veteran and past president of the American Physical Society—found himself repeating DeWitt's ratio of "two to three times" to journalists and in radio addresses without knowing (as his handwritten notes on typewritten speeches indicate) from whence the number had come or how it had been computed. Eager journalists soaked it all up.[15]

Lawmakers and their physicist consultants used the

Figure 7.1. Members of President Eisenhower's new Science Advisory Committee leaving the White House in October 1957. *From left to right*: David Z. Beckler, Isidor I. Rabi, Jerome B. Wiesner, and Charles Shutt. (*Source*: Photograph by Paul Schutzer, The *Life* Picture Collection, courtesy of Getty Images.)

launch of *Sputnik* and the purported "manpower gap" in science and engineering to push through the massive National Defense Education Act, signed into law in September 1958. The act authorized about $1 billion in federal spending on education (nearly $9 billion in today's currency), restricted to critical "defense" fields of science, mathematics, engineering, and area studies. The act represented the first significant federal aid to education in a century: not since the Morrill Land-Grant Colleges Act of 1862 had the federal government intervened so directly in higher education, which had traditionally been considered the prerogative of state and local governments. One close observer of the legislative wrangling behind the National Defense Education Act concluded that opportunistic policymakers had used the *Sputnik* scare and the DeWitt and Korol reports as a "Trojan horse": the act's proponents had been "willing to strain the evidence to establish a new policy."[16]

Passing legislation is usually a messy affair. The effects, in this case, were crystal clear. On the eve of the bill's passage, US institutions had been producing only 2,500 PhDs per year across all of engineering, mathematics, and the physical sciences. During its first four years, the National Defense Education Act supported 7,000 new graduate fellowships, about 1,750 per year. The huge federal outlay, in other words, amounted to an overnight increase of 70 percent in the nation's funding capacity to train graduate students in the physical sciences. During that same period, the act funded half a million undergraduate fellowships as well as providing block grants to institutions, with added incentives to states to increase science enrollments.[17] Hence the final element of economist Shiller's model: feedback.

:::

The feedback loop had an immediate impact on graduate-level training. During the decade after passage of the National Defense Education Act, the number of US institutions granting PhD degrees in physics doubled, driving an exponential rise in the number of young physicists. According to data collected by the National Register of Scientific and Technical Personnel—created during the early 1950s to facilitate the federal government's mobilization of scientists in times of war—the number of professional physicists employed in the United States grew substantially faster than the number of professionals in any other field between the mid-1950s and 1970: 210 percent faster than

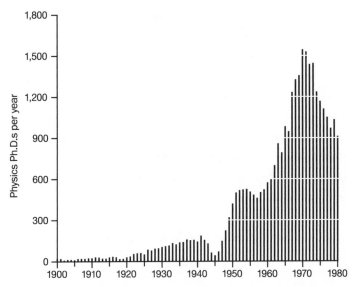

Figure 7.2. Number of physics PhDs granted per year by US institutions, 1900–1980. (*Source*: Figure by Alex Wellerstein, based on data from the US National Science Foundation.)

Earth scientists, 34 percent faster than chemists, 22 percent faster than mathematicians, and so on.[18]

Yet it was not to last. The conferral of physics PhDs in the United States peaked in 1971. Then, all at once, the curve fell sharply, its rate of descent an eerie reflection of its spectacular rise. A "perfect storm" had triggered the fall. By the late 1960s, internal auditors at the Department of Defense had begun to question whether the postwar policy of funding basic research on university campuses—which had underwritten the education of nearly all physics graduate students since the end of the Second World War—had produced an adequate return on investment. As the Vietnam War raged, meanwhile, campus protesters grew equally dissatisfied with the Pentagon's presence on American campuses; protesters often targeted physicists' facilities for their military ties (whether real or merely perceived). To supply troops for the escalation of fighting, military planners began to revoke draft deferments—first for undergraduates in 1967, then for graduate students two years later—reversing a twenty-year policy that had kept science students in their classrooms. Détente with the Soviets and the onset of "stagflation" in the early 1970s exacerbated the situation, inducing steep cuts in federal spending for defense and education.[19]

No field of study was affected more quickly or more dramatically than physics. While annual conferrals of PhDs across all fields slid by a modest 8 percent between their peak in the early 1970s and 1980, physics PhDs plummeted by half. Several fields experienced sharp downturns—mathematics down 42 percent, history down 39 percent, chemistry down 31 percent, engineering down 30 percent from the early 1970s highs—but physics led the way. Demand for young physicists vanished even more quickly. Whereas more employers than student applicants had registered with the Placement Service of the American Institute of Physics throughout the 1950s and into the mid-1960s, by 1968 young physicists looking for jobs outnumbered advertised positions by nearly four to one—and that included all kinds of positions, in academia, industry, and government laboratories. Three years later, the numbers were even grimmer: 1,053 physicist job seekers registered, competing for just 53 jobs.[20]

Robert Shiller is quick to note that speculative bubbles can develop without outright chicanery. That was certainly the case here. The influential physicists who used the DeWitt and Korol reports to argue for increased graduate training were doing their job; it was their responsibility to lobby on behalf of the profession. More generally, increasing support for higher education is hardly an evil thing. Yet in the rush to exploit DeWitt's "two to three times" finding, the cycle of hype, amplification, and feedback came unmoored from any reasonable assessment of the underlying situation.

Even if one set aside the significant caveats that DeWitt and Korol had delineated with such care—uneven quality of training, severe specialization, and the substantial numbers of extension and correspondence students inflating the Soviet statistics—the numbers themselves deserved a closer look. In tabulating numbers of graduates in engineering and applied sciences in the Soviet Union and United States, DeWitt had included three main categories: engineering, agricultural specialists, and health fields; these were the fields which, when tallied, produced the "two to three times" ratio. (As DeWitt explained, the grouping of fields in his tables was an artifact of the Soviet educational system, in which the vast majority of students earned degrees from "technical institutes" rather than universities; the institutes focused largely on these areas, whereas natural sciences and mathematics were mostly taught at universities.) Yet when repeating the "two to three times" number, not one commentator stopped to ask how a superabundance of agricultural specialists might lead to military supremacy—least of all given the Soviet Union's catastrophic history of collective farming in the 1930s or the fierceness with which the state had backed agronomist Trofim Lysenko's eccentric biological theories after the Second World War, quashing the study of genetics. Similarly for health professionals: no doubt an important segment of the labor force, but would more nurses and dentists lead to better bombs? Nearly everyone who picked up DeWitt's numbers used the label "science and engineering," never pausing to consider just which fields of science or engineering they represented.[21]

DeWitt had, in fact, included data on graduation rates in the natural sciences and mathematics for both countries, presented just as clearly as the information on agricultural and health specialists had been. Playing the numbers game with these data produced a rather different picture. Up through the mid-1950s (and, indeed, into the early 1960s, as DeWitt's follow-up study found), the United States maintained a two-to-one *lead* over the Soviet Union, rather than a deficit, in numbers of students who completed degrees in science and math each year. Lumping science, mathematics, and engineering graduates together and dropping agriculture and health, the ratio came out as 4:3 in favor of the Soviets—a lead that included a large fraction of students who earned their diplomas armed with a textbook and a mailbox. So much for the Soviets' "two to three times" advantage.[22]

None of these points were hidden in classified reports, sealed in a CIA safe; all were as plain on the page as the "two to three times" data had been. Amid a drumbeat of news about the Korean War, the launch of *Sputnik*, and nuclear brinksmanship, however, certain calculations proved persuasive. The sober rationality of economic analysis—so often honored in the breach—had succumbed to exuberance.

:::

The physicists' bubble, so sharply pronounced between 1945 and 1975, was not a one-shot deal. In fact, graduate enrollments in physics within the United States rebounded during the 1980s, bid higher and higher by many of the same

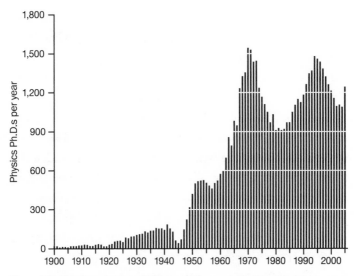

Figure 7.3. Number of physics PhDs granted per year by US institutions, 1900–2005. (*Source*: Figure by Alex Wellerstein, based on data from the US National Science Foundation.)

mechanisms that had inflated the first bubble. A resurgence of defense-related spending under the Reagan administration—including the sprawling Strategic Defense Initiative, or "Star Wars" program—combined with new fears of economic competition from Japan drove enrollments in physics and neighboring fields up exponentially once more, nearly matching the late-1960s peak. They fell sharply a decade later with the end of the Cold War. Just as during the early 1970s, shared conditions across fields led to an overall decline in graduate-level enrollments. By the time PhD conferrals in the United States bottomed out in 2002, annual numbers of PhDs granted across all fields had fallen by more than 6 percent from their 1990s peak. As before, though, some fields fell more sharply than others. Annual numbers of PhDs granted across all of science and engineering fell by 10 percent, while annual numbers of PhDs in physics plummeted by 26 percent. Once again, dire predictions of shortfalls in the scientific labor supply had been stupendously mistaken; once again, physics marked the extremes of a general pattern throughout American universities.[23]

The dynamics behind the second bubble were remarkably similar to the earlier example. Beginning in 1986, officials at the National Science Foundation had sounded the alarm again that the United States would soon face a devastating shortage of scientists and engineers. Foundation projections indicated that there would be 675,000 too few scientists and engineers in the United States by the year 2010. Just as in response to the DeWitt and Korol studies from the 1950s—especially the stripped-down ratio of "two to three times more" science and engineering graduates per year in the Soviet Union than in the United States—the dramatic projection of shortages during the 1980s helped to unleash generous federal spending.[24]

Unlike the DeWitt and Korol studies, the 1980s study by the National Science Foundation did not impress many close observers. In keeping with broader economic modeling during the Reagan administration, the study had neglected to consider demand at all, sticking only with supply-side variables. Yet few skeptics came forward until the early 1990s, after the Soviet Union had dissolved and the Cold War ground to an unexpected halt.[25]

Just as in the earlier era, reality checks that could easily have been applied were not, while the scarcity talk looped from hype to amplification to feedback all over again. And just as in the early 1970s, the second bubble burst, triggering double-digit unemployment rates among PhD-level scientists and mathematicians across the United States. The glut of freshly minted scholars—many of whom had been lured to graduate school with federally funded fellowships and promises of plentiful academic jobs to come—occasioned testy hearings in Congress. The pushback ultimately led to the dismantling of the Policy Research and Analysis Division within the National Science Foundation, which had developed the faulty supply-side projections.[26]

I witnessed this second bubble—or, rather, its sudden bursting—up close, having entered graduate school in 1993. Though I was just beginning my studies, I watched nervously as slightly older peers tried to navigate their way into academic careers. Whereas just a few years earlier there had been several faculty positions advertised each year across various subfields of physics, suddenly the best-trained students found themselves competing for a single position in a given specialty, often at an out-of-the-way university. A few years into my graduate studies, as the academic job market became ever more bleak, every single member of the latest batch of PhDs in particle physics from my department decamped for Wall Street, taking jobs as "quants" for the financial industry.[27] (My sister had recently begun working for a Wall Street firm and she, too, encouraged me to join her. "They love physicists here," she explained. "You can work on derivatives." When I told her that I worked on derivatives every day—confusing my calculus-laden homework problems with exotic creations like "collateralized debt obligations" that she had in mind—she just rolled her eyes.) Little could those young physicists know that they were fleeing one bubble only to help provoke another.

8

Training Quantum Mechanics

In the autumn of 1961, Richard Feynman launched a new experiment. Together with several colleagues at Caltech, he aimed to overhaul the curriculum for physics students. Their main goal was to introduce students to some of the most exciting—yet abstruse—aspects of modern physics as early as possible, right in their first year as undergraduates. That way, they hoped, they could fire the young students' imaginations, rather than making them wade through important but staid topics first. The capstone of the new syllabus, filling the final third of the yearlong course, centered on quantum theory.[1]

Feynman and his colleagues, Robert Leighton and Matthew Sands, feverishly composed the new lectures. Feynman delivered each one with his usual gusto, after which Leighton and Sands transcribed the recordings. Before long, rumors of the new course reached several textbook publishers. Feynman and his colleagues got to name their own terms. Leighton drew up a form letter, instructing interested publishers to submit written proposals to the authors within three weeks—a reversal of the normal procedure, in which authors submitted proposals to the publishers! The publishers were to describe how quickly they would be able to produce the books, at what price the new textbooks would be sold, what share of the royalties would be paid to Caltech, and what additional expenses the publishers proposed to absorb.[2]

In the end, Feynman and his colleagues chose to work with Addison-Wesley. Before the books came out, a sales representative took galleys on a tour to gauge interest among other physics faculty. Writing to the president of the press, the sales rep could hardly contain himself. "Comments: Great enthusiasm," began his long memo. "Where? In *every* department of physics, of course." Several faculty seemed to be amazed by the new book. "It took me a life time to leave his room *with* the Feynman book, he just wanted to read another chapter and another one!!" Another professor tried to brush him off, until the crafty salesman flashed the book's red covers. "Well . . . we had a nice talk for fifteen minutes and made an appointment for next spring. Of course he wanted a copy of the book." And so it went, town after town during the sales representative's two-week tour. "Give me a Feynman once or twice a year and I will do my job!" he closed. "I do not know who signed up Feynman, but I suggest that you owe him (not Feynman) a fine Turkey for his Christmas dinner!"[3]

The sales rep's instincts proved accurate. *The Feynman Lectures on Physics* sold more than 130,000 copies within six years of publication—even though Feynman himself later conceded that the pedagogical experiment had been a bit too ambitious. Some of the material really did prove to be

Figure 8.1. Richard Feynman lectures before a large undergraduate class at Caltech, ca. 1956. Recordings of lectures like these formed the basis for *The Feynman Lectures on Physics*, first published in 1963–65. (*Source*: Courtesy of the Archives, California Institute of Technology. Used with permission of the Melanie Jackson Agency, LLC.)

too advanced for first-year undergraduates. Yet sales remained brisk—indeed, the books remain in print today—driven largely by demand from more advanced students, and even faculty, who have continued to snatch up copies for self-study.[4]

Though Feynman, Leighton, and Sands might have gotten a bit ahead of the curve, by the early 1960s most of their colleagues shared their impulse to teach quantum theory to younger and younger students. Given how slowly university curricula usually evolve, the changes typified by *The Feynman Lectures* were extraordinary. Just twenty years before Feynman began lecturing on quantum theory to first-year undergraduates, many physicists in the United States had earned their PhDs without taking a single course in the subject.[5]

Amid the rapid-fire changes, some expressed alarm that too much was changing too quickly. A professor at Vanderbilt University, calling himself "one of those old-fashioned persons," suggested that "children should eat a reasonably good meal before partaking of dessert"—and, for this instructor at least, quantum theory was "distinctly dessert," which "could easily cause intellectual indigestion if not preceded by a properly balanced diet."[6]

Others, like J. Robert Oppenheimer, observed a more subtle shift: not just in what was being taught but how. Ever since the earliest work on quantum theory by the likes of Einstein and Schrödinger, Heisenberg and Dirac, the subject had inspired heated debate. So many of its core notions—the uncertainty principle, Schrödinger's cat, quantum entanglement—seemed to be at odds with other leading physical theories, let alone common sense. Yet when Oppenheimer surveyed how his colleagues taught quantum theory just a few years before Feynman, Leighton, and Sands launched their new course, he noted that the subject was by then "taught not as history, not as a great adventure in human understanding, but as a piece of knowledge, as a set of techniques, as a scientific discipline to be used by the student in understanding and exploring new phenomena."[7]

Quantum mechanics had become "an instrument of the scientist to be taken for granted by him, to be used by him, to be taught as a mode of action, as we teach our children to spell and add."[7]

How much things had changed. Oppenheimer had been among the first to bring working knowledge of the still-new quantum theory back to the United States, following his studies in Europe in the mid-1920s; before long, his course on the subject at Berkeley had become legendary. Yet by the time most of his colleagues began to offer their own courses on the subject, after the Second World War, the dramas of the wartime projects and the ensuing hyperinflation of physics enrollments had affected nearly every aspect of young physicists' training. The transition that Oppenheimer noted, perhaps a bit wistfully, in the 1950s—teaching quantum mechanics more as a toolkit than an adventure—became emblematic of broader shifts in the field after the war. Facing runaway enrollments, many physicists across the United States winnowed the range of what would count as "quantum mechanics" in the classroom. Where once-fabled teachers like Oppenheimer had relished talking through thorny conceptual challenges with small groups of students, instructors after the war—their intimate classrooms by then replaced by large lecture halls, tiered rows of seats teeming with students—increasingly aimed to train quantum *mechanics*: skilled calculators of the atomic domain.

:::

Oppenheimer's own entry into physics had been meteoric. Born in 1904 to a family of wealthy Jewish immigrants in New York City, he skipped several grades during his secondary schooling and entered Harvard for his undergraduate studies. (He later described his young self as "an unctuous, repulsively good little boy.") At Harvard he piled on extra courses each semester, graduating in just three years. During his first year as an undergraduate, he was invited to skip the introductory physics courses and dive directly into doctoral-level coursework.[8] As was typical at the time for the best American students who were interested in theoretical physics, Oppenheimer next set off for Europe to pursue his PhD, studying first at Cambridge before transferring to Göttingen. There he studied under Max Born, just as Born was collaborating with Werner Heisenberg and others to craft the new quantum mechanics and working frantically to try to make sense of the strange new formalism. Oppenheimer absorbed the emerging material quickly, publishing a dozen research articles while in Göttingen. He completed his PhD in the spring of 1927, a month shy of his twenty-third birthday.[9]

After a brief postdoctoral fellowship, Oppenheimer accepted teaching appointments in 1929 at both the University of California at Berkeley and at Caltech—more than 370 miles apart. Each department was so eager to hire him that they reached a compromise: Oppenheimer would teach at Berkeley in the fall and then decamp to Caltech for the winter and spring terms. During his first semester at Berkeley, he taught an elective course for graduate students on quantum mechanics. One student registered for credit, while twenty-five signed on to listen. During that first course, Oppenheimer raced through the material so quickly that students complained to the department chair; Oppenheimer grumbled, in turn, that he had to crawl so slowly through the syllabus. Before long, however, he developed an engaging lecturing style.[10] Graduate students routinely sat through his Berkeley course on quantum mechanics more than once; one desperate student staged a hunger strike until Oppenheimer relented and allowed her to attend the class for a fourth time.[11]

As late as 1939—the year that one of Oppenheimer's graduate students transcribed the lectures and made hectographed copies, which quickly saw wide circulation—Oppenheimer still introduced quantum mechanics as a "radical solution" to problems that were as much philosophical as physical. Lecture after lecture, he focused not only on the new mathematical formalism, centered on Schrödinger's wave function, ψ, but also on its curious physical interpretation. He lingered over Born's interpretation of $|\psi|^2$ as yielding probabilities for various outcomes, emphasizing the remarkable conceptual break from the rigid determinism of classical physics. Wielding Newton's laws or even Einstein's relativity, physicists had long since been able to calculate that B strictly followed A. In the newer world of quantum theory, on the other hand, physicists could calculate only likelihoods: B had certain odds to follow A, and physicists remained utterly stymied from saying more. Oppenheimer even indulged in Einstein-styled attempts to circumvent Heisenberg's uncertainty principle—revealing, with a flourish each time, how such clever efforts were destined to fail—all before walking his students through the first practical calculations with the formalism.[12]

Oppenheimer's pedagogical approach was hardly unique at the time. Felix Bloch, a Jewish émigré from Switzerland who had studied with Heisenberg before fleeing Nazism in 1934, taught his graduate-level course on quantum mechanics at Stanford University in a remarkably similar way. Throughout the 1930s, meanwhile, Caltech graduate students faced tough questions about the interpretation of quantum mechanics on their qualifying examinations. For years, beginning in 1929, the Caltech students kept communal notebooks in which they recorded how they had prepared for their oral exams and what questions various examiners had posed. Well into the late 1930s, faculty had pressed students to talk "all about [the] ψ function, physical meaning, etc." or had asked, "What is [the] interpretation of $\psi(x)$? Does the Schrödinger equation describe the rate of change for *all* time?"—a subtle question about how the range of probabilities encoded in the wave function reduces to a single, measured result. Then came the follow-up: "Discuss the nature of observation in quantum mechanics and in classical mechanics."[13]

The first textbooks on quantum mechanics by physicists in the United States likewise emphasized in their opening pages that students would need to confront "philosophical difficulties," which could not be "exorcised." Some even paused, in the midst of what would soon become a standard calculation of an electron's energy levels within a hydrogen atom, to assess whether various mathematical solutions could be considered physically meaningful if no experiment could distinguish between them. Others included entire chapters with titles like "Observation and Interpretation." Reviewers of the textbooks during the 1930s agreed that an overtly philosophical register was appropriate when it came to teaching quantum mechanics. They often disagreed with specific points of interpretation in the books under review, but not with the notion that textbooks should broach such interpretive issues.[14]

:::

Soon after the war, as more physics departments across the country began offering courses on quantum mechanics, the style of instruction began to shift. Few instructors during the 1950s lingered over how best to interpret the uncertainty principle or the place of probabilities in the quantum-mechanical formalism. Fewer still paused to dissect the philosophical standing of various hydrogenic wave functions.

The changes came on quickly. Some Caltech students, having studied reports of earlier oral exams, were caught by surprise. One complained in 1953 that the effort he had "invested in analysis of paradoxes and queer logical points was of no use in the exam." Instead, he had faced "straightforward questions" about then-standard calculations. Others similarly advised their fellow students to "give the usual spiel" or the "standard response" when asked to perform various quantum-mechanical calculations. One student suggested that his peers should simply "memorize" and "rehearse" answers to what had by then emerged as the standard calculations. Across the country, graduate students experienced a similar shift. Expansive essay-style questions about matters of interpretation, which had been common as late as the 1940s on the written qualifying exams, from Stanford and Berkeley to the Universities of Chicago and Pennsylvania, Columbia University, and MIT, were replaced by the mid-1950s by a collection of standard calculations.[15]

The pedagogical shift was closely correlated with enrollment patterns. Whereas Oppenheimer had lectured to about two dozen students at a time in his Berkeley course during the 1930s, after the war enrollments rapidly began to rise. By the mid-1950s, courses on quantum mechanics aimed at first-year graduate students typically had forty to sixty students enrolled; in the nation's largest departments, at Berkeley and MIT, the number edged over one hundred, "a disgrace [that] should not be tolerated at any respectable university," Berkeley's department chair complained to the dean. In a handful of departments, however, enrollments in first-year graduate-level quantum mechanics courses remained fairly small during the early 1950s or began to grow only later. Reading through lecture notes from these various departments reveals some remarkable differences. In short, an increase by a factor of three in enrollments was correlated with a decrease by a factor of five in the proportion of time devoted to the conceptual puzzles or philosophical challenges of quantum theory.[16]

Beyond the numbers and statistics, the lecture notes themselves provide some stark contrasts. Consider the course that Lothar Nordheim taught at Duke University in the spring of 1950. Like so many of his colleagues across the country, Nordheim had spent the war years working on the Manhattan Project. He had served as section chief at the Oak Ridge laboratory in Tennessee (principal site for isolating the fissionable isotope of uranium, U-235), rising to direct its physics division between 1945 and 1947. He left Oak Ridge for Duke in 1947 but did not stay long: by autumn 1950, he had begun to work full-time on the top-secret hydrogen-bomb project and later chaired the theoretical physics division at the major nuclear-related defense contractor General Atomics. In short, Nordheim was no stranger to the new realities of the military-industrial complex, and he excelled at wringing practical results, often under extreme time pressures, from the equations of quantum theory.[17]

Yet when he taught his course on quantum mechanics at Duke in 1950, Nordheim insisted that his students focus on its conceptual challenges. Working with a small class of a dozen students, he launched into the stubborn strangeness of quantum mechanics in his very first lecture. Given the new restriction to probabilities, he asked his students:

"What does this do to causality?" A student recorded in his notes, simply, "Ans[wer]. It Fucks it!" To drive the point home, Nordheim devoted two more lectures to the fabled double-slit experiment—a favorite example that Heisenberg and Schrödinger had each introduced in his own teaching, back in the early days of quantum theory, to emphasize such quintessential quantum features as wave-particle duality, superposition, and the uncertainty principle. Likewise for Nordheim's treatment of a quantum particle tunneling through a barrier. As he described the counterintuitive process, he pressed his students, "it is meaningless to ask, 'Is there causality?,' because we can never know the state completely at any time, because of [the] uncertainty relation. Hence, we discard the classical physical ideas of idealized observations."[18]

In other classrooms across the country, physicists who had shared many of Nordheim's worldly experiences—the secret, massive wartime projects, major consulting for defense projects after the war—charted a very different course when lecturing on quantum mechanics to their own graduate students. Where Nordheim lectured to a dozen students, most of these others faced classes that had already grown several times larger. At Chicago, Enrico Fermi spent twice as long deriving properties of the Laguerre polynomials— mathematical functions that quantify the behavior of an electron in a hydrogen atom—as he did on Heisenberg's uncertainty principle. At Cornell, Hans Bethe observed, with one passing remark, that trying to circumvent the uncertainty principle was as fruitless as designing perpetual motion machines, full stop. Even Richard Feynman, full of exuberance about bringing quantum theory to younger and younger students, made clear in his own classroom that the real purpose was to learn to calculate. In the lecture notes from his graduate-level course on quantum mechanics, he admonished that interpretive issues—of the sort that had filled Oppenheimer's lectures before the war and Nordheim's lectures after it—were all "in the nature of philosophical questions. They are not necessary for the further development of physics." While Nordheim had paused to consider conceptual sticking points of quantum tunneling, Freeman Dyson, lecturing to a class at Cornell with nearly three times as many students as Nordheim's class at Duke, plowed forward, adapting the usual calculation to treat various states of nuclear matter, such as deuterons. Dyson made clear, in his first lecture, that he would not follow the chosen textbook very closely. "Too much philosophy."[19]

:::

Two well-known textbooks, both published soon after the war, further illustrate the trend: Leonard Schiff's *Quantum Mechanics* (1949) and David Bohm's *Quantum Theory* (1951). Schiff and Bohm had each studied with Oppenheimer in Berkeley during the 1930s; both authors acknowledged how influential Oppenheimer's course had been for their own teaching. Yet what seemed like complementary models for teaching the subject—remarkably different in their emphases, yet equally hailed as great successes upon publication—soon collapsed under the pressure of rising student numbers.[20]

Leonard Schiff had been a postdoc with Oppenheimer between 1937 and 1940. He later joined the faculty at Stanford, and his *Quantum Mechanics* first appeared in 1949 to rave reviews. Schiff's book exemplified the toolkit approach to quantum mechanics. Whereas Oppenheimer had made his way slowly to the details of the Schrödinger equation, pausing at length to entertain many of the conceptual quandaries that arose along the way, Schiff largely dispensed with such philosophical niceties. ("We shall discuss physics, *not* philosophy," he announced on the first day of one of his courses in 1959.)[21] What had occupied nearly 20 percent of Oppenheimer's lecture notes, Schiff dispatched in a few opening pages of his book. In its place, Schiff provided what was widely hailed as the best collection of homework problems to calculate, of just the right level of difficulty for his target readers.[22]

David Bohm completed his PhD under Oppenheimer's direction in 1942 and published his *Quantum Theory* in 1951 after teaching at Princeton for a few years. He had tested out the material for his book in classes during 1947 and 1948, before Princeton's physics department had swelled too large. (His enrollments in those years were around twenty students in each class, similar in size to Oppenheimer's course at Berkeley in the 1930s.) Like Schiff's book, Bohm's book received glowing reviews at first—"a rare example of expressive, clear scientific writing," proclaimed one satisfied reviewer. In contrast to Schiff's approach, Bohm devoted several opening chapters to the kinds of philosophical challenges and conceptual puzzles that Oppenheimer, too, had emphasized. The Schrödinger equation didn't even appear until page 191 in Bohm's book; Schiff had first introduced the equation on page 21.[23]

The conceptual care that Bohm had taken when composing his textbook impressed several of his earliest reviewers. One praised "the concise and well balanced interplay, point-counterpoint, between formalism and interpretation." Another compared Bohm's and Schiff's books side by side—Schiff's being the only obvious American competitor published since the war—and offered the following balance sheet. Though only two-thirds as long, Schiff's book treated many more applications of the formalism in greater detail. Yet for those topics treated by both authors, this reviewer continued, "it is to the credit of Bohm's book that, for example, it gives the clearer and more physically understandable explanation."[24]

Despite their equally promising starts, the two books—like their authors—suffered quite different fates. Schiff became department head at Stanford and soon editor of the influential textbook series published by McGraw-Hill in which his own book had appeared. Bohm, meanwhile, was forced from his position at Princeton—and soon forced out of the country—just months after his book had been published. He had refused to name names when subpoenaed to testify before the House Un-American Activities Committee, during its headline-grabbing investigation into alleged "Communist infiltration" of the wartime Manhattan Project. Bohm fled to Brazil—where, in between crippling bouts of nausea, he was compelled to forfeit his US passport—before moving a few years later to Israel, eventually settling in London. Schiff's book saw two widely heralded, updated editions (in 1955 and 1968); Bohm's book was never reissued during his lifetime, and his efforts to publish a follow-up textbook on quantum mechanics were rebuffed.[25]

It fell to a third veteran of Oppenheimer's Berkeley group, Edward Gerjuoy, to make sense of the diverging paths. He took up the comparison in a review of the second edition of Schiff's book, in the mid-1950s. In expanding his book, Schiff had devoted even less space to conceptual or interpretive discussion; to Gerjuoy's taste, each edition of Schiff's book devoted too little attention to "such questions as correspondence, uncertainty, complementarity, and causality"—precisely the topics that had filled so much of Bohm's book. (Gerjuoy noted that "the contrast with Bohm's

Quantum Theory is interesting, even amusing.") But Gerjuoy could understand Schiff's decision not to amplify these topics in his revised edition. "With these subjects lecturing is of little avail—the baffled student hardly knows what to write down, and what notes he does take are almost certain to horrify the instructor, who perspicaciously usually resolutely refuses to question his students on these topics." So, instead, Schiff focused on a cache of worked examples: too soon, the student "is well into detailed algebraic complexities verifying which, he readily persuades himself to believe, means he is learning quantum mechanics." Though Gerjuoy could understand Schiff's pedagogical choices, he wondered—perhaps thinking back to his experiences as a student in Oppenheimer's famous course at Berkeley— whether it was "necessary, as Schiff does, to leap so rapidly over the philosophical issues raised by quantum mechanics that the student never has a chance to gauge their depth."[26]

Despite Gerjuoy's cautions, Schiff's textbook rapidly became the standard-bearer, its collection of homework problems especially well geared to teaching large classes of students. When asked to evaluate whether a third edition of Schiff's book would be warranted, a professor at Berkeley responded with a sixteen-page memorandum on why the previous two editions had been so successful. "I believe that the explanation is that Schiff is a very *practical* book," the reviewer began. "The reader who goes through the book really obtains a working knowledge of quantum mechanics." A student using the book, this reviewer continued, is "taken through a number of well chosen applications, and he is shown, through these examples how it all works out." It was an approach that the Berkeley physicist could appreciate; he had learned the subject from the first edition of Schiff's book. "As a student I was perfectly happy with this mode of presentation, and the book kept me sufficiently busy to prevent pseudo-philosophical speculations about the True Meaning of quantum mechanics."[27]

Many other physicists across the United States offered similar appraisals. Where once reviewers had evaluated textbooks on quantum mechanics at least in part on the basis of their philosophical stance, reviewers throughout the 1950s and 1960s routinely praised the latest offerings for "avoid[ing] philosophical discussion" and for omitting "philosophically tainted questions" that distracted from the business of learning to calculate. Enough with the "musty atavistic to-do about position and momentum," stormed MIT's Herman Feshbach.[28]

The new approach shaped the contents of the books as well. Between 1949 and 1979, physicists in the United States published thirty-three textbooks on quantum mechanics aimed at first-year graduate students. Together, these books included 6,261 homework problems (including, of course, many duplicate problems that appeared in several books). Most required students to manipulate the equations in the text: make a change of variables in the Schrödinger equation or evaluate various integrals. Only about 10 percent of the problems pressed students to go beyond the equations, to discuss their calculations in words. The pattern alarmed at least some older physicists, who, like Oppenheimer, had witnessed the remarkable conceptual transformations of quantum theory firsthand. In the early 1960s, one grumbled that with the spate of new textbooks, his colleagues had confused what was "easy to teach"—the "technical mathematical aspects of the theory"—with the conceptual understanding that students needed most.[29]

After the enrollments had crashed, however, newer textbooks began to appear, with a markedly different mix of homework problems. For example, Robert Eisberg and Robert Resnick pulled together a draft of their massive book, *Quantum Physics of Atoms, Molecules, Solids, Nuclei, and Particles*, in the early 1970s. By the time their book was published in 1974, first-year graduate enrollments in physics—the population to whom the book was directed—had fallen more than 60 percent from their 1960s peak. Eisberg and Resnick's book reflected the new classroom realities. In addition to hundreds of quantitative problems, akin to the classics that filled all three editions of Leonard Schiff's book, Eisberg and Resnick also included long lists of "discussion questions" at the end of each chapter. "Does a blackbody always appear black? Explain the term blackbody" was one early example. "What is the fallacy in the following statement? 'Since a particle cannot be detected while tunneling through a barrier, it is senseless to say that the process actually happens'"—hearkening back to one of Lothar Nordheim's favorite examples from his course at Duke. In a similar way, more than half of the homework problems within *Quantum States of Atoms, Molecules, and Solids* (1976), written by a trio of physicists at Rice University, were of this qualitative, essay-type form.[30]

:::

During the early 1950s, a young theoretical physicist at Berkeley learned the hard way how bloated class sizes could affect research and teaching styles. Having been on the faculty in Berkeley's physics department for a year and a half, the theorist was let go, not because he was unproductive in research or unconscientious in his teaching—the department chair insisted that the young professor had performed more than adequately at both. Rather, the theorist's chosen research topic fit poorly with the new pedagogical realities.

He had focused on rather abstruse points in quantum field theory. Though the topic could well prove important—the department chair considered it too early to say—it had failed an important test. Junior faculty members, the chair explained, needed to select research topics for themselves that could provide appropriate spin-off projects for their graduate students: "subjects that are not trivial, but at the same time are not unduly difficult or too time-consuming." Whether or not the young physicist's research would pan out in the long term, "it is not the sort of work that can readily be used for Ph.D. theses." With more than two hundred graduate students enrolled, Berkeley's physics department needed "someone who will be more useful to us." Only recently, in fact, the department chair had fast-tracked the promotion case for a different junior faculty member largely on the basis of his ability to craft appropriate problems for his many graduate students.[31]

Though few departments swelled as large or as quickly as Berkeley's, most felt the strain of the postwar enrollment boom. At nearby Stanford, physics faculty had prided themselves on the small-group intimacy their department could offer, compared with the "factory" at Berkeley. During the early 1950s, when the incoming cohorts included ten to twelve new students each year, Stanford faculty kept detailed notes on how individual students fared on their oral exams, the standard gateway between coursework and dissertation research: "Rather limited knowledge; shy, hesitant in answers; nervous," for example, or "well composed and thinks on his feet." Yet as the number of incoming students rose—soon up to thirty per year by the late 1950s, peaking at thirty-seven in 1969—the individualized note-taking stopped. The written exams shifted from essays to problems to calculate; faculty even flirted with administering true-false exams, to keep the burden of grading under control.[32] Physicists at the University of Illinois, facing similar pressures, jokingly called for a "test-ban treaty" in 1963—between faculty and students rather than the United States and Soviet Union—while students there lobbied for a "flunk-out shelter."[33]

Then the bottom fell out: only eighteen graduate students entered Stanford's department in 1970, and sixteen in 1972. Just as suddenly, the department once again undertook a sweeping reform of its comprehensive exams, restoring "a significant fraction of essay and discussion questions." In September 1972, the revised exam featured short-answer or essay questions in 40 percent of the problems, nearly double the proportion in the previous decade's exams. That same year, the department introduced a new, informal seminar on "speculations in physics"—just the sort of thing that had cost the young theorist at Berkeley his position twenty years earlier.[34] Richard Feynman took similar advantage of the transformed pedagogical realities at Caltech. He began to offer an informal course known as "Physics X," open to any undergraduates who were eager to puzzle through juicy scientific questions. One of my favorite photographs shows Feynman gesturing at the blackboard in 1976—the suit and tie from his early-1960s *Feynman Lectures* days replaced by an open, wide collar—while a handful of students look on, some sporting headbands, feet propped up on a desk.[35]

There has never been one "best" way to teach quantum mechanics. In particular, the enrollment-driven pragmatism, so stark in American physics departments after the Second World War, was anything but a "dumbing down." The second and third editions of Leonard Schiff's acclaimed textbook, for example, contained homework problems aimed

Figure 8.2. Richard Feynman teaching his informal "Physics X" course at Caltech in 1976. (*Source*: Photograph by Floyd Clark, courtesy of the Archives, California Institute of Technology. Used with permission of the Melanie Jackson Agency, LLC.)

at entry-level graduate students that would have stumped leading physicists only a decade or two earlier. Yet that tremendous accumulation of calculating skill came with some unnoticed trade-offs. For every additional calculation of baroque complexity that physics students learned to tackle during the 1950s and 1960s, they spent correspondingly less time puzzling through what those fancy equations meant—what they implied about our understanding of the quantum world.[36] Different ideals—about quantum theory, about what it meant to be a physicist—flourished while enrollments bulged, and after they went bust.

9

Zen and the Art of Textbook Publishing

Some books become totems, icons of an age. Stumbling upon them in an attic or used-book store, their crackling spines and musty air can trigger a rush of memories and associations—not just of where we were but *who* we were when we first encountered them. For hundreds of thousands of readers across the world, Fritjof Capra's *The Tao of Physics* continues to fulfill that role. First published in 1975, Capra's curious little book became a breakout publishing sensation. The book's unexpected commercial success inspired a wave of copycat books, reviving a dormant genre in popular-science writing about the mysteries of quantum theory. Its main argument—that modern physics had recaptured, even recapitulated, the age-old wisdom of the Eastern mystics—wasn't exactly new; several creators of quantum theory had made similar pronouncements back in the 1920s and 1930s. But unlike the long-since-forgotten analogies proffered by Niels Bohr, Erwin Schrödinger, and their colleagues, Capra's paperback achieved mass appeal. For a generation of countercultural seekers, the book promised a union of Western science and New Age enthusiasms.

Capra's *Tao of Physics* reveals some of the larger ways in which hard-nosed science became enmeshed with countercultural delights during the long 1970s, blurring boundaries that might otherwise have seemed distinct. The conditions of the book's composition and the variety of uses to which it was put clarify broader patterns in the entanglement of academic physics with its publics after the war—and how, for a brief moment at least, physics became groovy.[1]

To make sense of Capra's book and the roles it came to play, we must look back to 1945, not just to 1968 or 1975. The booming enrollments in physics classrooms during the 1950s and 1960s had hastened changes in the ways physicists broached subjects like quantum mechanics. Yet the fantastic growth in student numbers proved unsustainable, and the sudden reversal of classroom conditions— after the crash of the early 1970s—opened space again for the return of a speculative or interpretive idiom. The material that helped to fill that pedagogical void was often inflected by the growing New Age and counterculture movements, just then gathering steam across North American university campuses. Some of the iconic books from the era began to operate in multiple registers: both popular books for the masses and textbooks for science students. Capra's *Tao of Physics*—the most emblematic and successful of these efforts—exemplifies the hybrid nature of the new books and the diverse roles they came to play.[2]

:::

The book, like its author, traveled a long route to American classrooms. Austrian-born Capra completed his PhD in theoretical particle physics at the University of Vienna in 1966 and moved on to a postdoctoral fellowship in Paris. There the student uprisings and general strikes of May 1968 left a deep impression on him. He also met a senior physicist from the University of California at Santa Cruz who was spending some sabbatical time in Paris. The professor invited Capra to Santa Cruz for a follow-up postdoctoral fellowship, which Capra gladly accepted. He arrived in Santa Cruz in September 1968.[3]

Capra broadened his horizons on many fronts in California. As he later wrote, he led "a somewhat schizophrenic life" in Santa Cruz: hardworking quantum physicist by day, tuned-in hippie by night. He continued his political education, already stoked by the Paris of 1968: he went to lectures and rallies by the Black Panthers; he protested against the war in Vietnam. He took in "the rock festivals, the psychedelics, the new sexual freedom, the communal living" that had become de rigueur among the Santa Cruz counterculture set. He also began exploring Eastern spiritual traditions—an interest originally sparked by his filmmaker brother—by reading essays and attending lectures by Alan Watts, a local expert on Buddhism, Hinduism, and Taoism.[4]

In the midst of these explorations, Capra had a powerful experience on the beach at Santa Cruz during the summer of 1969. Watching the ocean waves roll in and out, he fell into a kind of trance. As he later described it, the physical processes all around him took on a new immediacy: the vibrations of atoms and molecules in the sand, rocks, and water; the showers of high-energy cosmic rays striking the atmosphere from outer space; all these were more than the formulas and graphs he had studied in the classroom. He felt them in a new, visceral way. They were, he gleaned, the Dance of Shiva from Hindu mythology. Inspired by his experience on the beach, he soon noticed similar parallels between cutting-edge quantum theory and central tenets of Eastern thought: the emphasis upon wholeness or interconnectedness, for example, or upon dynamic interactions rather than static entities.[5]

In December 1970, his visa about to expire, Capra returned to Europe. With no new job lined up, he began to check with some of his contacts to see if he might find some steady position. He wandered into the theoretical physics division at London's Imperial College, whose leader he had met in California. The physicist had no fellowships to offer—finances had become as difficult for British physicists as for their American colleagues by that time—but with the financial downturn there were at least some empty desks around. And so Capra set up shop at Imperial: no position, no income, but a tiny corner of office space he could call his own.[6]

His financial situation quickly grew dim. He took on private tutoring jobs; he did some freelance work writing abstracts of recent physics articles for the *Physikalische Berichte*. When he could spare the time, he delved more deeply into his readings of Eastern texts, inspired as much by Alan Watts's teachings as by his own mystical experience on the beach. And he hatched a plan to put some of his hard-won physics knowledge to use: he would write a textbook on his beloved subject of quantum physics. If he could write the book quickly enough, he reasoned—and if he could get a major textbook publisher interested in the project—he might pull out of his financial tailspin. Not only that, the textbook might make him a more attractive candidate for a teaching position down the road.[7]

By November 1972 he had drawn up an outline for the book and begun drafting chapters. He reached out to another contact for advice: MIT's Victor Weisskopf, whom he had met at a recent summer school in Italy. Weisskopf, like Capra a native of Vienna, was by that time a grand old man of the profession. He had recently completed a term as director general of CERN, the multinational high-energy physics laboratory in Geneva. By the time Capra sought Weisskopf's advice, the elder physicist was well into a sideline career as a successful popular-science writer. He had also published a highly influential textbook on nuclear physics—a book that held the honor, Weisskopf was always happy to recount, of having been most frequently stolen from the MIT libraries. Weisskopf had suggested the idea to Capra of writing a textbook when they met at the summer school. Capra sent his chapter outline to Weisskopf, hoping for some further encouragement. He also hoped that his senior colleague would use his contacts to help line up a publisher and secure an advance payment in anticipation of future royalties.[8]

Back and forth their letters flew: Weisskopf commenting on Capra's proposal and soon on individual chapter drafts; Capra thanking him for his comments but pressing again and again for more tangible forms of support. "As you know, the problem of financial support has become vital for me," Capra responded in January 1973, "and I wonder whether I could approach a publisher for a contract" at that stage of the project. If so, which publisher would Weisskopf recommend, and would Weisskopf mind contacting the press directly to recommend the book? "I am sorry to bother you with these problems, but I have indeed very little time to work on the book at the moment, because I am not supported by anybody and have to make my living with much less creative work." Weisskopf's responses—asking for more drafts and sending along further comments—sidestepped the issues Capra found most pressing. Capra reiterated his urgent need to line up a publisher and get some financial support.[9]

A few weeks (and chapter drafts) later, Weisskopf addressed Capra's main concern. "I like your style and find many things well expressed," he began. "I would again like to encourage you to go ahead and finish the manuscript." But, Weisskopf advised, Capra should wait before approaching a publisher until he had a complete manuscript in hand. He should also understand that few publishers offered advances for textbooks any more. "I understand your need for financial support but I suppose you are aware of the fact that a book like this is not going to bring in much money because of the nature of the subject. The best that one can hope is something like $1 thousand the first year and less thereafter." Writing a textbook, Weisskopf counseled, might be a noble endeavor, but it was a lousy get-rich-quick scheme.[10]

Just at that moment, Capra received an invitation to visit Berkeley and give some talks to physicist Geoffrey Chew's group. (Capra had sent some early essays comparing Chew's core notion of particle physics—a self-consistent particle "bootstrap"—to central doctrines of Buddhist thought. Chew had passed these essays on to two graduate students, who in turn encouraged Chew to invite Capra to visit.) While back in California, Capra also checked in with his former postdoctoral adviser at Santa Cruz. They talked about Capra's parallel projects: continuing his exploration of Eastern spiritual traditions and pushing forward on his textbook. The Santa Cruz physicist—"a rather hard-headed and pragmatic physicist" in Capra's estimation, hardly one drawn to the woolly countercultural currents swirling around him—encouraged Capra to combine his interests and change gears with his book project. Rather than write a physics textbook, why not refocus the book to explore the parallels between physics and Eastern thought that had so intrigued Capra since his transcendental experience on the beach? Coming on the heels of Weisskopf's realistic cautions about how well a textbook might sell, Capra took his former adviser's advice. Upon his return to London, Capra began composing new chapters on Eastern traditions—one each on Hinduism, Buddhism, Confucianism, Taoism, and Zen—and interleaving them with the textbook chapters he had already written.[11]

Capra found the new plan inspiring and set about trying to interest publishers in the project. A dozen rejections later, a small London-based publishing house agreed to take a gamble on it, even offering Capra the long-sought, if modest, advance payment that allowed him to finish writing it up. Completed manuscript in hand, Capra next managed to interest a tiny American publisher to bring out an edition in the United States: Shambhala Press, then just five years old, which had been founded in Berkeley to publish books on Eastern mysticism and spirituality. *The Tao of Physics* thus appeared simultaneously in Britain and the United States in 1975.[12]

A few months later, Capra presented Weisskopf with a copy of the book, when they were both attending a conference in California. Weisskopf read most of it on his plane flight back to Massachusetts, and, as he reported back, he "liked it very much." "It is very hard for me to judge whether you have succeeded in your task," Weisskopf continued, "since it addresses itself to a very specific kind of public than you find here in the East." (Translation: we have no hippies at MIT.) "I do believe, however, that it is a good book and that there will be many people who will have a better idea of physics after they have read it." Weisskopf shared his concern that some readers might be "scared off by the 'Tao' side of the deliberation" but conceded that "you can't make it right for everybody." He closed on a brighter note: "I wish you all luck and wonder how the sales will go."[13]

They went well. The first edition from Shambhala —20,000 copies—sold out in just over a year. Bantam brought out a pocket-sized edition in 1977 as part of its "New Age" series, with an initial printing of 150,000 copies. By 1983, half a million copies were in print, with additional editions prepared in a host of foreign languages. Twenty-five years later the book had achieved true blockbuster status: forty-three editions, including twenty-three translations—everything from German, Dutch, French, Portuguese, Greek, Romanian, Bulgarian, and Macedonian to Farsi, Hebrew, Chinese, Japanese, and Korean—with millions of copies sold worldwide.[14]

Many factors seem to have combined to launch the book into the sales stratosphere. For one thing, Capra enjoyed a firm command of the physics; he had been well trained. The fact that the physics-heavy portions of his book had begun as drafts for a textbook—and that those sections had benefited from careful readings by a towering physicist like Viki Weisskopf—surely helped Capra clarify just how he wanted to present difficult concepts such as Heisenberg's uncertainty principle and quantum nonlocality. Moreover, his incursions into Eastern thought, while sometimes belittled by specialists in religious studies, nonetheless sprang from a genuine earnestness.[15] Capra had become a seeker, reading everything he could get his hands on. By the time he finished the book, he had spent years experimenting with alternate modes of encountering the world, always pushing to absorb the insights of the ancient mystical traditions. And then there was his impeccable timing. With the New

Age rage in full force by the mid-1970s, conditions were ripe for a book like *The Tao of Physics*. Capra's book capitalized on a diffuse, widely shared craving to find some meaning in the universe that might transcend the mundane affairs of the here and now. The market for Capra's book had been teeming like a huge pot of water just on the verge of boiling. *The Tao of Physics* became a catalyst, triggering an enormous reaction.

:::

When Capra set out to promote the book, he seemed straight out of central casting. "Tall and slim with curly brown hair skirting the nape of his neck," cooed one *Washington Post* reporter. "Capra, with California tan, shoulder bag, and a Yin Yang button pinned to his casual jacket, seems more a purveyor of some new self-awareness scheme than a physicist." It soon became clear, however, that Capra was more than just a pretty face. He was on a mission not just to explore the foundations of modern physics but to alter the very fabric of Western civilization: "a cultural revolution in the true sense of the word," as he put it in the book's epilogue. As he saw it, modern physics had undergone a tremendous sea change in its understanding of reality, and yet most physicists—let alone the broader public—had failed to appreciate the consequences. The "mechanistic, fragmented world view" of classical physics had been toppled by quantum mechanics and relativity, but Western society still carried on as if Einstein, Bohr, Heisenberg, and Schrödinger had never lifted a pencil. "The world view implied by modern physics is inconsistent with our present society, which does not reflect the harmonious interrelatedness we observe in nature," he explained. A proper understanding of what modern physics had achieved—especially its "philosophical, cultural, and

Figure 9.1. Fritjof Capra discussing *The Tao of Physics* in November 1977. (*Source*: Photograph by Roger Ressmeyer, courtesy of Getty Images.)

spiritual implications"—could help restore the balance before it was too late.[16]

Capra's main argument throughout *The Tao of Physics* was that modern physicists had rediscovered the teachings of the age-old sutras of Buddhism, Vedas of Hinduism, and *I Ching* of ancient Chinese thought. "The further we penetrate into the submicroscopic world, the more we shall realize how the modern physicist, like the Eastern mystic, has come to see the world as a system of inseparable, interacting, and ever-moving components with man being an integral part of this system." Capra marched through a series of these parallels. First and foremost was what he saw as the "organicism" or holism implied by quantum interconnectedness: ultimately the quantum world is not divisible into separate parts but is woven into one seamless whole.[17] Capra also saw deep parallels between the koans, or riddles, of Buddhist thought, the constant interplay of opposites in Taoism, and the paradoxes of quantum theory. Niels Bohr's complementarity called on physicists to transcend what appeared to be opposites: neither wave nor particle but both. Although "this notion of complementarity has become an essential part of the way physicists think about nature," Capra explained, the physicists had come late to the party: "in fact, the notion of complementarity proved to be extremely useful 2,500 years ago," when the Chinese sages promoted the dialectic of yin and yang to the center of their cosmos. Little wonder, Capra concluded, that Bohr adopted the yin-yang symbol for his family coat of arms. Einstein's relativity, meanwhile, with its interconversion of matter and energy via $E = mc^2$, echoed the Eastern emphasis upon dynamism and flow: the universe caught in a never-ending dance rather than being a collection of static objects. The merging of space and time into a unified spacetime likewise brought the physicists' cosmic picture into line with long-standing Eastern intuitions.[18]

The Tao of Physics succeeded in that rare category, the crossover hit. It held broad appeal for hundreds of thousands of readers who were not physicists, or academics of any sort. Looking back, a few years after its original publication, one reviewer marveled that Capra's book had sold "amazingly well, not only to the usual Shambhala devotees of Eastern religion but also to engineers, Caltech grad students and people of the general population who, a few years later, would be reading Carl Sagan." Reviewers routinely touted Capra's clear expository style. In fact, the book received a significant amount of serious, scholarly attention. Academic journals specializing in philosophy, history, and sociology carried reviews. The journal *Theoria to Theory* published a lengthy review section on the book, with detailed comments from three specialists in philosophy and religious studies. Sociologists and philosophers of science likewise devoted substantial articles to the book, picking through the claimed parallels and subjecting each to sustained critique.[19]

Perhaps the most surprising response of all, however, came from scientists. Some certainly responded as we might expect, downplaying the book as mere popularization and dismissing the countercultural overtones as just so much *zeitgeistische* pap. Famed biochemist and science writer Isaac Asimov, for one, bewailed the "genuflections" to all things Eastern made by "rational minds who have lost their nerve." Jeremy Bernstein, Harvard-trained physicist and staff writer for the *New Yorker*, went further. He concluded his review of Capra's book, "I agree with Capra when he writes, 'Science does not need mysticism and mysticism does not need science but man needs both.' What no one needs, in my opinion, is this superficial and profoundly misleading book."[20]

These predictable responses, however, were by no means the norm. Mysticism aside, Capra offered a vision around which many physicists could rally. In his opening chapter he had noted the "widespread dissatisfaction" and "marked anti-scientific attitude" of so many people in the West, especially among the youth. "They tend to see science, and physics in particular, as an unimaginative, narrow-minded discipline which is responsible for all the evils of modern technology." "This book aims at improving the image of science," Capra declared; the insights and joys of modern physics extended far beyond mere technology. Indeed, "physics can be a path with a heart, a way to spiritual knowledge and self-realization." Few reviewers missed the point. *Physics Today* ran a review of the book by a Cornell astrophysicist. The review began by citing the profession's litany of woes: the "anti-scientific sentiment" of the age, which distressed Capra and his critics alike, "manifests itself on all levels of our society, from a decrease in funding for basic research to a turning to Eastern mysticism and various forms of occultism." Not an auspicious start for the volume under consideration. Yet the reviewer judged Capra's book to be a great success. For one thing, the book got the physics right. Even more important: *The Tao of Physics* integrated "the abstract, rational world view of science with the immediate, feeling-oriented vision of the mystic so attractive to many of our best students."[21]

The reviewer's comments proved more than a passing observation. Just as the review was going to press, Capra was busy teaching a new undergraduate course at Berkeley based on his book. He reported proudly to MIT's Victor Weisskopf that one-third of the students were science majors, eager to learn about the foundations of modern physics: just the sort of philosophical material they were not receiving in their other physics classes. Soon the *American Journal of Physics*, devoted to pedagogical innovations in the teaching of physics, began carrying articles on how best—not whether—to use *The Tao of Physics* in the class-room. One early adopter began by citing the huge market success of Capra's book. "This leads naturally to the question," he continued, "how can a physicist utilize this interest by offering a course using Capra's book?" A follow-up article commented matter-of-factly: "Anyone involved in physics education is likely to be asked to comment on parallelism [between modern physics and Eastern mysticism] at some stage. It would be easy to dismiss such ideas entirely, and in so doing possibly risk alienating a new-found interest among students. This field has the potential of appealing to the imagination and should perhaps be carefully explored and maybe even 'exploited.'" With budgets falling and en-rollments crashing, physicists could ill afford to turn their noses up at anything that might bring students back into their classrooms.[22]

As late as 1990, university physics courses throughout North America still routinely listed *The Tao of Physics* on their syllabi as a "helpful reference."[23] When critics com-plained that Capra's book might just as likely confuse stu-dents as enlighten them—why becloud difficult concepts from quantum physics with other difficult concepts from Eastern thought?—some of the physicists who had adopted Capra's book for classroom use were quick to respond. "It should be emphasized that most of these students would not have taken an offering in the Physics Department if it were not this one," came one spirited reply. His *Tao*-centered physics course had become one of the best in his depart-ment at delivering "bums in the seats."[24] Indeed, using Capra's book had inspired this physicist to develop new lesson plans on topics like Bell's inequality and quantum entanglement—that beguiling interconnectedness that

Einstein had dismissed as "spooky actions at a distance"—which still had not made their way into standard physics curricula or textbooks.[25]

:::

In a roundabout way, Capra thus fulfilled his original goal: he wrote a successful textbook after all. Physicists across the continent eagerly snatched up *The Tao of Physics*. In their classrooms, the book helped demonstrate to disaffected students—or so physicists hoped—that physicists, too, were "with it."[26] Capra's surprising commercial success with *The Tao of Physics* presaged a wave of similar books, including Gary Zukav's *The Dancing Wu Li Masters* (1979), Fred Alan Wolf's *Taking the Quantum Leap* (1982), and Nick Herbert's *Quantum Reality* (1985). Like Capra's punchy paperback, these books sold handsomely, and several netted national awards. They also discovered second lives in physicists' classrooms. Just as reviewers had done for *Tao*, physicists touted the popularizations as useful textbook proxies. Next-generation textbooks on quantum mechanics—composed after the crash in enrollments had become the new normal—quoted liberally from the popular books and recommended them for further reading.[27]

Books like *The Tao of Physics* capture a moment when all that seemed solid nearly melted into air. The boundaries between disciplined academic physics and an inchoate countercultural youth movement, on the one hand, and between peer-reviewed textbooks and blockbuster best sellers, on the other, assumed a newfound plasticity. No single arrow pointed from one domain to the other; no diffusion or osmosis smuggled bits of "real" physics to the New Age seekers or back the other way. Instead, smart and well-trained young scientists—earnest in their pursuit of wide-ranging questions, caught up in a tectonic shift of professional roles and expectations, and immersed in the counterculture's technicolor bloom—charted a new way to be a physicist during the tie-dyed 1970s.[28]

MATTER

10

Pipe Dreams

On 10 September 2008, I huddled with colleagues around a laptop, furiously clicking "refresh" on the web browser. We were watching real-time updates as the first batch of protons began their inaugural lap around the Large Hadron Collider, or LHC, a brand-new particle accelerator near Geneva. Revved up to enormous speeds by supercooled magnets, the protons raced around the LHC's huge ring, twenty-seven kilometers in circumference. They crisscrossed the French-Swiss border more than ten thousand times per second before smashing into each other, releasing primordial fireworks.

Watching the LHC come on line was a thrilling moment for me, but also a bittersweet one. Squinting at the tiny laptop screen, my thoughts began to wander. I imagined a similar celebration off in some parallel universe—a celebration that never was. For almost exactly fifteen years earlier, construction on a similar machine, even grander than the LHC, had ground unceremoniously to a halt. That other machine was known as the Superconducting Supercollider, or SSC; it was to be based outside Dallas, Texas, in the small town of Waxahachie. (The town's other main attraction: Southwestern Assemblies of God University.)

As an undergraduate, back in 1992, I had worked as an intern for a few months at the Lawrence Berkeley National Laboratory, in northern California. In Berkeley I joined a tiny subdivision of a sprawling, international collaboration that was building a huge detector to be used at the SSC.[1] I used to joke with friends that my term in Berkeley served as my "foreign study": having grown up on the East Coast, Berkeley seemed just as exotic and unfamiliar as the tales my friends told of their semesters in Edinburgh, Florence, and Tokyo. My first day on Berkeley's main campus, down the hill from the Lab, student protesters had hoisted themselves up to the top of the iconic Campanile, a three-hundred-foot-tall clock tower on campus. There they set up a rickety platform and refused to come down until all experiments on campus involving laboratory animals had stopped. And *that* was before I had discovered Telegraph Avenue. I might as well have moved to Mars.

I wrote a research article during that internship, predicting some features of the fleeting interactions among subatomic particles that the SSC was designed to observe. The first draft began confidently, with the matter-of-fact scientific prose that young students quickly learn to imitate: "The high energies and luminosities available when the Superconducting Supercollider (SSC) comes on-line have intensified interests in probing various extensions of the Standard Model." The eyes of a generation of physicists were focused on the SSC and on the riches it promised to reveal. I was eager to become one of them, a gadfly on the outskirts of billion-dollar science.

As my article bumped along through peer review, however, larger forces—far more powerful than the particle collisions that we anticipated within the SSC—began to play out. By the time I submitted a revised version of my article, the SSC's political fortunes had changed dramatically. I dropped all reference to the SSC, substituting a generic line about future generations of accelerators, off in the indefinite future.[2] Not long after that, the US Congress took its final vote to kill funding for the SSC. As it happens, that vote occurred about four weeks after I had begun my doctoral studies in high-energy physics. A few days before the final vote in Congress, a well-meaning young professor called me into his office. He advised me to leave graduate school if the vote went the wrong way. I stayed, but he didn't: a year or so later he jumped ship to Wall Street, along with so many other students and colleagues. With that single vote to end support for the SSC, Congress cut annual funding for high-energy physics in the United States in half. Support for the field continued to erode, losing ground against inflation, for the rest of the decade.

Since that time, scientists, policymakers, and historians have spilled much ink over what led to the SSC's demise. (Even novelist Herman Wouk, of *Caine Mutiny* fame, got into the act with his 2004 novel, *A Hole in Texas*.) Some point to cost overruns; others highlight deeply felt differences over how to distribute limited resources across the gamut of scientific research. All agree on one major factor: the Cold War had ended.[3]

:::

Figure 10.1. The partially excavated tunnel for the Superconducting Super-collider in Waxahachie, Texas, early 1993. In October 1993, Congress canceled the project. (*Source*: Fermilab Archives, SSC Collection.)

How different that moment was, in the early 1990s, from what had come before. Berkeley physicist Ernest Lawrence had helped to usher gigantism into American physics during the 1930s, building a series of larger and larger particle accelerators known as "cyclotrons" in the hills overlooking Berkeley's campus. The first models had fit easily on a table-top; later they swelled to fill rooms and eventually whole

Figure 10.2. Ernest Lawrence and his staff pose with the newly renovated 184-inch synchrocyclotron at the Berkeley Radiation Laboratory in 1946. (*Source*: Lawrence Berkeley National Laboratory.)

factories' worth of space. Lawrence could measure his progress by the size of his machines. His team graduated from a model with an 11-inch diameter in January 1932 to a 27-inch replacement that December. In the fall of 1937, Lawrence commanded a 37-inch cyclotron, which itself was eclipsed by a 60-inch machine less than two years later. By 1940, funding was in hand, and one thousand tons of concrete were poured to support his 184-inch cyclotron.[4]

In the late 1940s, the US secretary of war visited Lawrence's laboratory, which had been a major contracting site for the Manhattan Project during the Second World War. After giving the secretary a tour, the entrepreneurial Lawrence mentioned that he could really use some additional funding for his latest project. The secretary gamely assured Lawrence that the Army would be glad to support his efforts. Just before leaving, the secretary stopped to ask, casually, "By the way, Professor Lawrence, did you say two million or two billion?" Both figures seemed equally plausible.[5] By that time, Lawrence's Berkeley model had found emulators across the country, as the Office of Naval Research and the Atomic Energy Commission supported the construction of dozens of new particle accelerators and nuclear reactors across the country. None started out as tiny as Lawrence's originals had been. Rather, they chased after the lead that Lawrence continued to provide, such as his massive Bevatron, first operational in 1954.[6]

During those years, policymakers calculated that building such machines would be important for national security. The argument was not that the enormous devices would lead in any direct way to better bombs. Rather, in keeping with the view that scientific "manpower" was a vital national resource, the machines would serve as training devices with which to expand the pool of experts on whom the nation could call in an emergency. In 1948, for example, the Atomic Energy Commission agreed to fund not one but two massive accelerators, one at Lawrence's Berkeley laboratory and the other at a new facility on Long Island, New York, even though the scientific advisory group to the commission had argued that only one such machine could be justified on the basis of scientific merits. The decision to build both machines was made out of fear for the "morale" of the physicists at whichever laboratory had lost the bid. As a commissioner explained the following year, funding such machines would bring the government not just "big equipment" but "big groups of scientists who will take orders."[7] The connection became even more explicit after the United

States entered the Korean War. In July 1951, an official with the Atomic Energy Commission argued that the commission should build even more particle accelerators. He went through a simple calculation: if N nuclear physicists were "willing, able, and eager to use particle accelerators, and on average five such men per accelerator is an effective team," then the commission should build $N/5$ accelerators, or two per year for as long as "the international situation remains roughly as at present."[8]

The first hints of trouble—that the federal firehose of spending on enormous machines like particle accelerators might not last forever—surfaced in the late 1960s, ultimately triggering the dramatic nosedive in physicists' enrollment curve. In 1969, one of Lawrence's protégés had to defend a proposal to build yet another particle accelerator, even grander than the rest, in the Midwest. He parried pointed questions from Congress about high costs and pragmatic ends. The proposed facility, the physicist calmly responded, would not aid in the nation's defense; it would make the nation "worth defending."[9] The rhetoric worked, and the new accelerator was installed at Fermilab, outside Chicago, but only because the director's public testimony had been reinforced by concerted backroom lobbying by a coalition of scientists and policymakers. In 1983, during a resurgence of defense-related spending by the Reagan administration, that coalition scored another victory, winning endorsement from the administration to begin construction on a still-larger machine, the SSC. By the time the SSC's number came up, though, ten years later, that coalition had dissolved. Despite similarly soaring rhetoric from Nobel laureates, who promised Congress that the SSC would unlock the secrets of the universe and contribute to an epic adventure of discovery, their words fell flat.[10] By the early

1990s, with no Soviet menace to face (real or imagined), the blank-check era of American big science had come to an end.

:::

A year after the SSC was abandoned, the governing board of CERN, the European Organization for Nuclear Research in Geneva, approved its own plan to pursue the LHC. CERN's leaders realized they could achieve goals similar to the SSC's, but on the cheap. Most important, they decided to use a preexisting tunnel from an older experiment to save on the huge excavation costs. The choice was not ideal—using the older facility meant sticking with a colliding ring that was only one-third as long as the SSC's would have been. The size of the ring had a direct bearing on the energies that the colliding particles could attain; these, too, would be only one-third as high in the LHC as they would have been in the SSC. But the CERN machine nevertheless could reach enormous energies at about one-fifth the cost of the SSC. And so we celebrated that day back in September 2008—fourteen years after the commitment had been made—when the LHC spun to life and sent its first batch of protons circling around and around and around.

Operation of the machine came screeching to a halt just a few days later. Faulty electrical connections deep inside the LHC tunnel caused several magnets to overheat. That, in turn, led one of the tanks holding liquid helium to rupture. (The liquid helium was needed to keep the superconducting magnets ultracold.) No one could get close to the affected area to inspect the damage or begin repairs until the entire region had been taken off-line and ever so slowly warmed up. After fourteen months and an additional expenditure of nearly $40 million, the tank had been repaired, new equipment installed to try to bolster the LHC's resistance to similar spikes in electrical current, and the entire machine cooled back down to its operating temperature. A new batch of protons were sent on their dizzying journey, round and round inside the huge accelerator.[11]

Late in November 2009, the laboratory team celebrated a new world record: they had achieved the highest-energy particle interactions ever recorded in an Earth-bound accelerator, edging past the previous record set by the smaller machine at Fermilab. Even that record-setting energy, however, remained roughly ten times lower than the anticipated peak energy for which the LHC had been designed. Once again bitter disappointment quickly eclipsed the momentary cheer. The lab announced in spring 2010 that it would be able to operate the LHC at only half capacity until the end of 2011, before taking the entire machine off-line for a new round of costly repairs. The culprit again appeared to be electrical shielding around the delicate superconducting magnets. A few years later—in an encounter that Aesop surely would have relished—a tiny weasel shut the monstrous machine down after scaling a fence, gnawing through some electrical cables, and receiving a jolt of 18,000 volts.[12]

Weasels notwithstanding, the LHC has flourished at CERN while the unfinished SSC recedes further into memory. When Congress halted funding for the SSC in October 1993, the federal government had already spent $2 billion on the project and excavated nearly fifteen miles of underground tunnel; Congress had to appropriate another $1 billion just to cover shutdown costs. The Cold War model, which had once seemed so self-evident to scientists and policymakers—that huge research projects would be funded to ensure that generations of scientists would be well trained and at the ready, in case the Cold War ever turned hot—had at last run its course. CERN, on the other hand, had been founded in 1954 on a rather different premise. Its purpose was to bring together scientists from many countries across Europe, despite at-times considerable political differences, to provide a platform for international collaboration. From the start, projects at CERN have been shielded from any whiff of potential military relevance, let alone classified research.[13] It was hardly obvious, back in the early 1950s, that European governments would pay such large sums to build enormous particle accelerators—but once that commitment had been made, the arguments undergirding CERN's efforts proved more adaptable to the political realities of a post-Soviet world.

Today, even multinational projects like a next-generation LHC face enormous hurdles, given the colossal price tags required to smash particles together at still-higher energies. And so particle physicists' hopes focus, at least for now, on the frozen vacuum of the LHC, buried deep underground, as they wonder whether the massive machine will be the last of its kind.

11

Something for Nothing

Just after New Year's Day, early in 1964, Murray Gell-Mann submitted a short paper to the journal *Physics Letters* that forever changed physicists' lexicon. He had been trying to make sense of curious patterns in the masses and interactions of dozens of newly discovered particles. In his brief, two-page article, he suggested that these exotic nuclear particles—as well as more familiar particles, like protons and neutrons—might themselves consist of smaller particles. Lifting a nonsense word from James Joyce's *Finnegans Wake*, Gell-Mann called the new particles "quarks." (Joyce's novel appears as Ref. 6 in Gell-Mann's brief reference list.) At nearly the same time, George Zweig at CERN wrote a lengthy paper introducing the same basic idea; he called the hypothetical entities "aces." Gell-Mann's paper was published within three weeks of submission; five years later, the Nobel Prize committee cited the work when awarding Gell-Mann the physics prize. Meanwhile, Zweig's paper was re-

Figure 11.1. Murray Gell-Mann introduced the term "quark" into physicists' lexicon early in 1964 and helped build particle physicists' Standard Model of elementary particles and interactions. (*Source*: AIP Emilio Segrè Visual Archives, *Physics Today* Collection.)

jected; neither that paper nor a longer, follow-up study ever made it into print. And so it was that physicists around the world added the jabberwockian term "quark" to their everyday speech.[1]

Ten years after Gell-Mann pirated Joyce, physicists had developed a full-fledged theory of how quarks behave. More than that: they had learned how to combine the new description of quarks with other, fresh ideas about the particles and forces that roil the nuclear realm. A product of many authors, the new coagulation of ideas became known simply as the "Standard Model"—a far cry from Gell-Mann's whimsical and idiosyncratic terminology, reflecting instead its composition by committee. The Standard Model describes the forces and interactions among all the known subatomic particles, from quarks and electrons to their most exotic cousins, which are seen only in carefully controlled experiments. For nearly half a century, the model has served as both compass and polestar, enabling physicists to navigate complicated experimental results and guiding new inquiries. The Standard Model has been subjected to high-precision tests since the 1980s; nearly all the tests have shown exquisite agreement with predictions. (The only tantalizing discrepancy to date concerns neutrinos, whose tiny but nonzero masses had not been incorporated into the original model.) Even the long-elusive Higgs boson—the last hypothetical particle described by the model to be detected—provided a close match to predictions. The Standard Model is almost certainly the most boringly titled exciting development in the history of science.[2]

For all its successes, however, physicists agree that the Standard Model cannot be the final word. For one thing, it suffers from a series of arbitrary, unexplained features. Why does one particle, the muon—otherwise so similar to the electron—happen to weigh precisely 206.7683 times more than its lightweight sibling? Why do two particular interactions have strengths in the ratio 0.23120 rather than, say, 1 or 0.25 or 17? By sticking those parameters in by hand, physicists can match experiments to extraordinary accuracy. But accounting for why those values must be put in, just so, remains an open question. For several decades, high-energy physicists have striven to account for those parameters in a first-principles sort of way—that is, to develop some larger framework into which the Standard Model might be subsumed, in terms of which the arbitrary features might appear natural, even necessary. Beyond the arbitrary parameters, meanwhile, most physicists consider the Standard Model glaringly incomplete. It incorporates three of the four basic forces of nature: the forces that cause electric charges to attract or repel; that cause nuclear particles to clump densely into atomic nuclei; and that cause some of those nuclei to disintegrate via radioactivity. But the Standard Model has nothing at all to say about gravity, which, on cosmic scales, is by far the most important force of all.

Most of the efforts to redress these shortcomings—to rectify the arbitrariness of the existing Standard Model and to glean some way to smuggle gravity in—focus on symmetry. Symmetry means that a system remains unchanged even as you shake it up or twist it around or, as physicists say, perform a transformation. Imagine playing a Bach fugue on the piano. Unbeknownst to you, some mischievous gremlins have shifted your keyboard up by a major third. Every time you play what looks like middle C, the piano key's hammer actually strikes the strings for E; you hit a D but the hammer sounds an F-sharp, and so on. If the gremlins' actions affect each note in the same way, independent of their position on the keyboard—and if they don't change the rules over time—then they have performed a "global transformation." The relative intervals between notes have remained intact, but the piece is not truly unchanged. Someone with perfect pitch could detect the difference.

The fugue would remain unchanged—symmetric under this global transformation—if, equally unbeknownst to you, tiny elves living inside the piano hooked up an elaborate contraption of pulleys and gears. Every time a piano key's hammer begins to fall, the elves' wheelworks redirect it to the originally intended strings. By adding in more machinery—new types of particles and interactions—the elves compensate for the gremlins' transformation, leaving the composition completely unchanged from the original.

So much for symmetry under global transformations. More complicated transformations are possible, too. For example, the gremlins inside the piano might dream up a distinct transposition for every note on the keyboard: middle C moves up to E, while D moves down to B-flat, and so on. Even worse, the gremlins might change their minds and make up different transpositions over time, so that later the middle-C hammer strikes a G while the D hammer strikes a D-sharp. Physicists call such maneuvers "local transformations." With the right combination of gears and pulleys, the elves could still render your Bach fugue unchanged from the original, if the elves constantly adjusted their machinery to compensate for the gremlins' complicated transformations. In the parable, the elves' machinery enforces the symmetry. More than that: the whole reason to dream up the elves in the first place, with their pulleys and gears — to posit that the world really contains more kinds of stuff than we originally thought — is to protect the hypothetical symmetry. On this telling, thought-stuff — specific, mathematical symmetries — would conjure up physical things in the world, populating the subatomic realm with particular kinds of particles.[3]

The forces described by the Standard Model — forces that bind nuclei together or make them fall apart — remain symmetrical under local transformations. Decades ago, physicists postulated that special particles might exist that generate compensating nuclear forces which, just like the elves' impressive bric-a-brac, guarantee the overall symmetry. The types of transformations the particles needed to overcome helped to fix what properties they should have, if indeed

Figure 11.2. Carlo Rubbia and colleagues conducted a series of experiments at CERN, code-named UA1 and UA2 (named for "Underground Area"), that succeeded in finding evidence of the hypothetical W and Z force–carrying particles of the Standard Model in 1983. (*Source*: CERN, courtesy of AIP Emilio Segrè Visual Archives.)

they existed at all. And, lo and behold, when experimenters went looking for particles to match those descriptions in the early 1980s, using large particle accelerators at Fermilab, CERN, and elsewhere, there the particles were: pretty much exactly as the theorists had expected they would be.[4]

That was a remarkable congruence. Small hints gleaned from various experiments suggested that some underlying symmetry might govern specific forces of nature. Mental gymnastics—often far more elaborate, far-fetched, and just plain bizarre than my gremlin-elf-piano story—led theorists to predict that some new, tiny things might be out there scurrying around, shoring up the underlying symmetry. New experiments then aimed to catch a fleeting glimpse of those tiny elves at work, or at least to find empirical data that might plausibly be attributed to those dreamed-up interactions. The Standard Model was pieced together that way between the 1960s and the 1980s, a frenetic zigzag between theory and experiment.

One of the astonishing successes of the Standard Model has been to account for why objects have mass. The Higgs mechanism (to which I turn in the next chapter) describes a process by which basic constituents such as quarks and electrons acquire mass from their immersion in a vat of Higgs-field goo. But what about conglomerations of quarks, like protons and neutrons? Nearly all the matter we know— you, me, just about everything we can see here or in the heavens—consists of protons and neutrons. (There seems to be quite a lot of matter in outer space that we can't see, known as "dark matter," which is not composed of protons and neutrons, but let us tackle one cosmic mystery at a time.) Only about 5 percent of the mass of ordinary particles like protons and neutrons can be accounted for by the mass of their indwelling quarks. Ninety-five percent of a proton's mass—and, by extension, 95 percent of the mass of you and me—comes from raw energy. Mass does not arise from glomming lots of heavy items together. It comes very nearly from nothing: from the feverish quantum dance of massless particles.[5]

The main participants in this dance are gluons. Gluons are one species of elves from my Bach parable: they skitter around enforcing a particular symmetry. The symmetry they regulate governs the strong nuclear force, that is, the force that binds quarks together into composites like protons and neutrons. As their name implies, they are nuclear glue. Gluons do not have any mass of their own—they avoid all those jostlings with the Higgs field—but they interact with each other and with quarks all the time. Most important, they stay true to their elven ways. If you try to disturb the symmetry they guard—for example, by placing a lone quark in isolation—gluons leap into action, dredging up other quarks with compensating nuclear charges to cancel out the first quark's charge and restore the overall symmetry.

The cancellation would be complete if the compensating quarks could be forced to sit directly on top of the original quark; then no quark-charge would spill out to threaten the nuclear-force symmetry. But a competing factor hinders any complete cancellation. The Heisenberg uncertainty principle, that central pillar of quantum theory, stipulates a mandatory trade-off between how precisely a quantum object's position and momentum may be specified. In other words, nothing—not even gluons—can force quarks to sit perfectly still in a fixed location. The more gluons act to keep the new quarks affixed squarely on top of the original, the more energetically those quarks jump around, like so many toddlers pitching a tantrum. At the natural balancing point between those two tendencies—canceling the original quark's charge as much as possible while minimizing the new quarks' thrashing—some residual energy remains. Thanks to Einstein's $E = mc^2$, we see that energy in the form of a proton's mass. Our mass, you might say, is nothing but a cosmic accounting error.

The basic mechanism of this mass-inducing process had been hypothesized in the 1970s, just a decade after Gell-Mann had first dreamed up the idea of quarks. Yet it took years to evaluate the idea. For one thing, calculating the quantitative details of the quark-gluon interactions proved formidable. Only in 2008 could the first compelling computer simulations tackle realistic-enough scenarios to allow physicists to compare theoretical predictions for the mass of the proton with experimental data. The match was remarkable, the accounting secure. Those imagined elves really seem to be out there, performing their tasks just as the Standard Model prescribes.[6]

So here we are: teeming collections of atoms, which are mostly empty space, their subatomic constituents acquiring heft from the symmetry-preserving whirl of a gluonic quantum dance. "Standard Model," indeed.

12

Higgs Hunting

Particle physics is at once the most elegant and the most brutish of sciences. Elegant because of its sweeping symmetries and exquisite mathematical structures. Brutish because the principal means of acquiring information about the subatomic realm is by revving up tiny bits of matter to extraordinary energies and smashing them together. Richard Feynman once likened particle physicists' methods to trying to discern the inner workings of a finely crafted pocket watch—carefully gauged springs and gears, all arranged just so—by hurling two watches at each other and watching the detritus that comes flying apart.[1] In particle physics, there's an added twist: some of the detritus was never contained within the original matter. It's as if, in addition to the springs and gears of the smashed watches, out flew pulleys, ropes, the odd coin, and a yo-yo or two. The new objects that come flying out when subatomic particles smash together are coagulations of raw energy: some of the energy carried by the two colliding particles becomes transmuted, thanks to Einstein's famous equation, $E = mc^2$, into little chunks of matter. These colossal transmutations occur billions of times per second in hulking machines like the Large Hadron Collider (LHC) at the CERN laboratory near Geneva.

As soon as the LHC came on line, in September 2008, huge teams of experimentalists began looking for one particle in particular: the Higgs boson. ("Boson" is a generic label for particles that carry whole-number units of spin, or intrinsic angular momentum, such as a photon of light. Most of the particles that make up ordinary matter, such as protons and electrons, carry half-integer units of spin and are known as "fermions.") The Higgs has been dubbed "the God particle," though I have never understood why this particular bit of matter is presumed to be holier than all the others. I prefer a more descriptive nickname: the "billion-dollar boson," since for decades the challenge of finding the Higgs particle served as a major argument in favor of building larger and larger particle accelerators.[2]

The idea of the Higgs boson emerged more than fifty years ago, an idea born of desperation. Experiments had suggested that the weak nuclear force—responsible for phenomena such as radioactive decay—obeyed specific symmetries. Several theorists recognized that they could model such a force if, as for electromagnetism, the weak force arose when subatomic particles exchanged special force-carrying particles. But there was a catch. The symmetries of the weak force demanded that the hypothetical force-carrying particles have no mass, just like the massless photon that gives rise to electromagnetic forces. Unlike electromagnetism, however, the weak force has a very short range: it is effective only when particles are very close to each other (such as

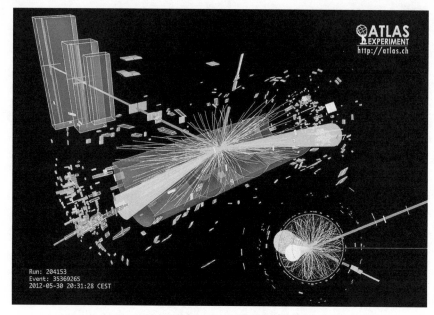

Figure 12.1. Reconstruction of particle paths from a single event captured in the ATLAS detector at CERN in late May 2012. Protons that had been accelerated to near light speed collided, forming a short-lived Higgs boson. Before it could leave a measurable track in the detector, the Higgs particle quickly decayed into two tau-mesons, which in turn decayed into an electron (thin line pointing nearly straight up from the collision region) and a muon (thin line pointing diagonally up and to the left). (*Source*: ATLAS Collaboration, courtesy of CERN.)

packed tightly within an atomic nucleus); the short range, in turn, seemed to imply that the force-carrying particles should be very massive. Hence the conundrum: theorists could model the symmetries of the weak force or its short range, but not both.[3]

Several physicists proposed a clever workaround. What if the force-carrying particles of the weak force really were massless but were always slogging through some medium—a medium that fills all of space and slows the force-carriers' motion, like marbles rolling through molasses? Several versions of that idea appeared in the journal *Physical Review*

Letters over the summer and fall of 1964, in short papers by François Englert and Robert Brout (received at the journal on 26 June and published on 31 August), by Peter Higgs (received on 31 August and published on 19 October), and by Gerald Guralnik, Carl Hagen, and Thomas Kibble (received on 12 October and published on 16 November). Higgs noted in his paper that the molasses-like medium implied that there should exist a new particle, associated with the medium, which became known as the "Higgs boson."[4]

On paper, the Higgs boson acquired a central role in what emerged as the Standard Model of particle physics. Though the hypothetical Higgs particle was the plainest of all the particles described by the model — zero electric charge, zero intrinsic angular momentum, no abstruse quantum properties such as "strangeness" or "color charge" — its function, to bestow mass on would-be massless particles, became critical. The Higgs would give the others their heft. Nobel laureate Frank Wilczek playfully dubbed the Higgs boson the "quantum of ubiquitous resistance"; CERN theorist John Ellis made an analogy to people trudging across a snow-covered meadow.[5]

The idea was compelling, but it remained only an idea for decades. As the years wore on, particle physicists learned that it was one thing to posit a universal medium through which all matter lumbers along and quite another to produce empirical evidence that such a medium exists — in essence, to break off tiny pieces of that medium (individual Higgs particles) and measure their properties.

The strategy to find evidence of Higgs particles seemed straightforward enough: smash particles like protons together at such high speeds that Higgs bosons (along with a great deal of other stuff) would coagulate from the residual energy. After search upon search came up empty, physicists were able to place limits on the Higgs boson's mass: if such objects really existed in nature, individual Higgs particles should weigh nearly as much as an atom of gold. Unlike gold atoms, however, physicists expected Higgs particles to be remarkably evanescent, with a lifetime of roughly a trillion-trillionth of a second. Such objects, physicists knew, wouldn't sit around long enough to be photographed, nor would they leave much of a track: the furthest they could travel before decaying into other particles would be about one ten-trillionth of a centimeter.[6]

The only hope of finding evidence of Higgs particles was to sift through their decay products, or the decay products of their decay products, trying to distinguish signals of a long-since-vanished Higgs boson amid all the other subatomic wreckage that streams away from the collision region in a particle accelerator. This is a bit like trying to infer the existence of a particular long-deceased grandmother—and measure her height and weight—by sifting all the data of a national census. One must look for statistical deviations from the expected patterns of ordinary particles that get trapped in the detectors. Are there more particles of a particular type, with particular energies, than would be expected in the absence of a Higgs boson that had decayed into them?

No "golden event," captured in a single photograph, can terminate such a search; no shriek of "Eureka!" marks its conclusion. Rather, by necessity, discovery claims depend upon complicated statistical arguments.[7] Physicists must collect terabytes of data from all the trillions of scatterings and interactions and stack them into histograms: graphs showing the numbers of events of a given type at a given energy. Then they must carefully subtract away the "background," or expected patterns of particle decays from

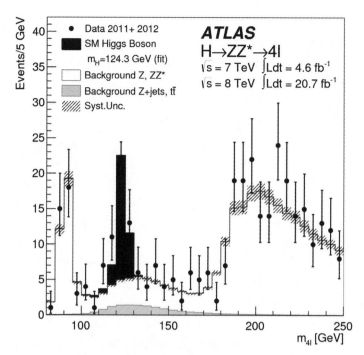

Figure 12.2. Statistical evidence of an excess number of decay events associated with a mass of around 125 GeV/c^2, which would be consistent with a Higgs boson decaying into a pair of Z bosons, which in turn decay into four leptons (such as electrons, muons, or neutrinos). The number of events at that particular mass was significantly higher than would be expected in the absence of a genuine Higgs boson with a mass of about 125 GeV/c^2, given the low level of other "background" events that should have been measured at that energy. (*Source*: ATLAS Collaboration, courtesy of CERN.)

known processes other than the creation and decay of a Higgs boson, and check whether any excess signal remains. (One of my favorite books in graduate school was *The Higgs Hunter's Guide*, filled with the arcana of calculating the expected signals and backgrounds for Higgs decay at various energies.[8] The title made me feel like Indiana Jones.) Higgs hunting is a game of looking for bumps in the night: tiny but otherwise unexplainable deviations in those histograms.

In December 2011, representatives from two large teams

at CERN convened a press conference to share evidence that they were on the trail of the elusive Higgs particle. But they had to stop shy of claiming a discovery. By that time, each group had collected and analyzed impressive amounts of data, culled from trillions of collisions among subatomic particles—but not enough to rule out statistical flukes.[9]

The teams at CERN needed to collect information from even more scatterings in order to clarify that the tiny bumps in their data were due to a new particle rather than to an unlikely string of events arising from ordinary, non-Higgs phenomena. Toss a coin ten times and it may well come up heads six times. In fact, you can expect to get six heads out of ten tosses about 20 percent of the time—a not-uncommon statistical fluke. Such a limited data set (only ten coin tosses) is not large enough to determine with confidence whether you are using a fair coin or a biased coin that favors heads over tails. But if you tossed the coin ten thousand times and got six thousand heads, that would point more convincingly toward some real effect—an inherent bias toward heads— rather than an ordinary coin on a lucky streak.

Years ago, particle physicists adopted the convention that discovery claims for new particles would require statistical significance of at least five standard deviations (usually denoted "five sigma"). That means that the odds that the observed events could have been due to mundane particles on a lucky streak—rather than arising from a genuinely new particle—are about three million to one. Neither of the teams at CERN that were searching for the Higgs boson had collected sufficient data to reach the five-sigma mark in December 2011. That changed a few months later. At a dramatic press conference on 4 July 2012, both teams announced that they had crossed the five-sigma mark with data collected through June. They had amassed clear evi-

Figure 12.3. ATLAS Collaboration spokesperson Fabiola Gianotti (*left*) and particle theorist Peter Higgs (*right*) congratulate each other at the press conference at CERN on 4 July 2012, soon after Gianotti presented her group's evidence that they had detected the long-elusive Higgs boson. (*Source*: Photograph by Denis Balibouse / AFP, courtesy of Getty Images.)

dence of the Higgs boson.[10] The following year, Peter Higgs and François Englert shared the Nobel Prize.

How can we take the measure of such an achievement? One can certainly focus on the money: billions of dollars spent on the ill-fated Superconducting Supercollider; billions more expended at CERN to build and maintain the LHC. Budget lines certainly provide an important measure—not least during times of financial hardship—but they are not the only ones to consider.

A different kind of accounting may help to explain physicists' exuberance at the CERN press conference on 4 July 2012. By that time, one of the standard databases of scientific publications included more than 16,000 articles on the Higgs particle, stretching back to the early articles from the 1960s.[11] More than 90 percent of those articles had been published since 1990, and nearly 1,000 had appeared in 2011 alone. Those 16,000 articles were written by about 11,000 authors: physicists around the world who had been focusing on the Higgs particle, its theoretical roles and possible experimental detection, for decades. Five hundred of those authors had each published at least 55 articles on the topic, dedicating a large portion of their careers to the Higgs particle. (Four of my papers showed up on that list, less than 0.03 percent of the global effort. John Ellis from CERN led the pack with 150 articles. When he compares the Higgs field to a snow-covered meadow, you can be sure he knows what he's talking about.)

Hence the jubilation that greeted the news from the LHC that July—a celebration not just for the five thousand or so physicists affiliated with the two teams at CERN but for the thousands more around the world who had contributed to the quest over half a century. Physicist Matthew Strassler declared 4 July 2012 to be "IndependHiggs Day."[12] I couldn't imagine a better reason for fireworks.

13

When Fields Collide

Sometimes I need to kick myself—*hard*—for overlooking things that were right under my nose. I had one of those experiences about a decade ago when I noticed a paper posted to the world's central electronic physics preprint server, arXiv.org. The authors of the new paper proposed a beautiful model, soon dubbed "Higgs inflation," that could account for specific phenomena during the very early universe. Their idea was elegant and simple: what if the Higgs particle of the Standard Model also had a particular, nonstandard coupling to gravity, of the sort that had been hypothesized (in a rather different context) several decades earlier. Then the selfsame Higgs boson, which particle physicists already needed as part of the Standard Model, might answer even larger questions about the structure and evolution of our universe as a whole.[1]

I kicked myself for not proposing the idea myself. After all, I had already published several papers exploring aspects of Higgs-like fields in the early universe, which incorporated the nonstandard gravitational interactions; in fact, the authors of the new paper cited some of my previous research.[2] Yet I had never taken the next step—a step that should have seemed obvious to me—and now they had. Stepping back from my initial frustration, the experience offered an opportunity to think about why certain questions come to seem obvious to various researchers, even if other scholars—trained in different ways, viewing the field from a different perch—had overlooked such questions altogether. In my own case, the new model fit squarely within my own specialty, a subfield of physics known as "particle cosmology."

Particle cosmology is flourishing these days. The field investigates the smallest units of matter and their role in determining the shape and fate of the entire universe. In recent years the field has received generous funding from governments and private foundations; these have supported state-of-the-art satellite missions and enormous ground-based telescopes, as well as underwriting the research of thousands of theoretical physicists around the world. An average of more than two new preprints on the topic are posted to arXiv.org *every hour* of every single day—nights, weekends, and holidays included.[3]

The field's dramatic success is all the more striking given that it barely existed forty years ago. The rapid rise of particle cosmology illustrates the potent alchemy of ideas and institutions that I find so fascinating—an entanglement often clearest in hindsight. In this case, the new subfield took form as some physicists aimed to push beyond the still-new Standard Model of particle physics; the newer ideas, emerging during the mid-1970s, took on special salience as unprecedented changes shook the discipline, especially in the United States. A generation before Congress killed the Superconducting Supercollider, particle physicists in the United States had faced a similar crisis. Their responses—institutional and curricular—enacted rapidly in the mid-1970s, helped to push certain questions to the research frontier, even as other research programs faltered.

These complex forces are thrown into starkest relief by following the fortunes of two sets of ideas: one introduced by gravitational specialists, the other puzzled over by particle physicists. Neither of these sets of ideas drove the union of particle physics and cosmology. Rather, tracing their fates over time clarifies larger processes. To unpack some of the ways that changes in politics and institutions can affect intellectual life, we may turn to the problem of mass.

:::

During the 1950s and 1960s, physicists in at least two subfields struggled to understand why objects have mass. Mass seems like such an obvious, innate property of matter that one might not even think it requires an explanation. Yet finding descriptions of the origin of mass that remained compatible with other ideas from modern physics proved no easy feat.[4] The problem took different forms. Experts on gravitation and cosmology framed the problem in terms of Mach's principle. Mach's principle—named for the physicist and philosopher Ernst Mach (1838–1916), famed critic of Newton and inspiration to the young Einstein— remains stubbornly difficult to formulate, but a good approximation might be phrased this way: are local inertial effects the result of distant gravitational interactions? In other words, does an object's mass—a measure of its resistance to changes in its motion—ultimately derive from that object's gravitational interactions with all the other matter in the universe? If so, do Einstein's gravitational field equations, the governing equations of general relativity, properly reflect this dependence?[5]

Within the much larger community of particle physicists, the problem of mass arose in a different form. Theorists struggled to incorporate masses for elementary particles without violating the symmetries that seemed to govern nuclear forces. Beginning in the 1950s, particle theorists faced a dilemma: either they could model the symmetries of these nuclear forces, which seemed to require the absurd step of setting all particles' masses to zero, or they could incorporate particle masses in their equations but destroy the symmetries.[6]

Around the same time, physicists in both specialties developed schemes to explain the origin of mass. Both proposals postulated that a new field existed in the universe, whose interactions with all other types of matter would account for why we see those objects as possessing mass. On the gravitation side, Princeton graduate student Carl Brans and his thesis adviser, Robert Dicke, pointed out in a 1961 article that in Einstein's general theory of relativity — by then physicists' reigning description of gravity — the strength of gravity was fixed once and for all by Newton's constant, G. According to Einstein, G had the same value on Earth as it did in the most distant galaxies; its value was the same today as it had been billions of years ago. Brans and Dicke suggested, instead, that Mach's principle could be satisfied if the strength of gravity varied over time and space. To make this variation concrete, they hypothesized that some new, physical field φ permeated all of space, taking on different values here versus there, now versus then. The new field fixed the force of gravity: $G \sim 1/\varphi$, so that G would now vary inversely as φ did. (In regions of space in which φ had a

Figure 13.1. *Top*, Carl Brans during his graduate studies at Princeton, 1959. (*Source*: Courtesy of Carl Brans.) *Bottom*, Brans's dissertation adviser at Princeton, Robert Dicke. (*Source*: Photograph by Mitchell Valentine, courtesy of AIP Emilio Segrè Visual Archives, *Physics Today* Collection.)

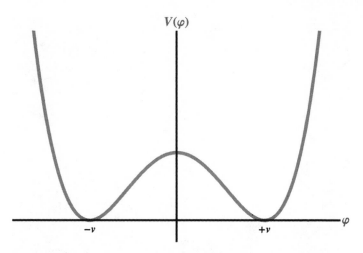

Figure 13.2. A double-well potential, $V(\varphi)$. The energy of the system has a minimum when the field reaches either of the values $+v$ or $-v$. Although the field's potential energy is symmetric, the field's solution will pick out only one of these two minima, breaking the symmetry of the governing equations. (*Source*: Illustration by the author.)

large value, G would be small, and vice versa.) They swapped $1/\varphi$ for G throughout Einstein's gravitational equations. In the resulting model, ordinary matter would respond *both* to the curvature of space and time, as in ordinary general relativity, *and* to variations in the local strength of gravity, coming from φ. All matter interacted with φ, and hence the new field's behavior helped to determine how ordinary matter moved through space and time. Any measurement of an object's mass would therefore depend on the local value of φ. Brans and Dicke's idea seemed so compelling that members of the gravity research group at Caltech used to joke that they believed in Einstein's general theory of relativity on Mondays, Wednesdays, and Fridays, and in Brans-Dicke gravity on Tuesdays, Thursdays, and Saturdays. (On Sundays they relaxed at the beach.)[7]

Several specialists in particle physics attacked the problem of mass around the same time, also introducing a new, hypothetical field that might pervade the universe. Jeffrey Goldstone, for example, observed in 1961 that the solutions to various equations need not respect the same symmetries that the equations themselves do. As a simple illustration he introduced a new field, which he also labeled φ. The potential energy for his new field had two minima, one at a value of $-v$ for the field φ and the other at the value $+v$.

The energy of the system is lowest at these minima, and hence the field will eventually settle into one of these values. The potential energy is exactly the same—symmetric—for each of these values of the field, even though the field must eventually land in only one of them. This ingenious idea quickly became known as "spontaneous symmetry breaking": whereas the curve of potential energy is fully left-right symmetric, any given solution for φ would be concentrated only on the left or on the right.[8]

A few years later, in 1964, Peter Higgs revisited Goldstone's work. He found that when he inserted Goldstone's idea into models of the highly symmetric nuclear forces, spontaneous symmetry breaking would yield *massive* particles. In such models, the new field φ would interact with other particles, including the force-carrying particles that generated nuclear forces. The equations governing these interactions, Higgs demonstrated, obeyed all the requisite symmetries. Before φ settled into one of the minima of its potential, these other particles would skip lightly along, merrily unencumbered. Once the φ field arrived at either $+v$ or $-v$, however, it would exert a drag on anything interacting with it—like all those marbles mired in molasses. Once that happened, the subatomic particles would behave as if they had some nonzero mass; any measurements of their mass, in turn, would depend on the local value of φ.[9]

Figure 13.3. *Top*, Jeffrey Goldstone identified the idea of spontaneous symmetry breaking in 1961. (*Source*: AIP Emilio Segrè Visual Archives, *Physics Today* Collection.) *Bottom*, Peter Higgs incorporated Goldstone's symmetry-breaking idea into a model of nuclear forces in 1964. (*Source*: Photograph by Robert Palmer, courtesy of AIP Emilio Segrè Visual Archives.)

Both sets of papers — by Brans and Dicke and by Higgs — received immediate attention from colleagues. One helpful measure of their impact comes from counting citations. Over the years, physicists have refined the art of citation counting, tallying the number of other scientific articles that refer to, or cite, a particular article in question. The standard citation-tracking database for high-energy physics assigns papers to various categories on the basis of their accumulated citations. Papers that have never received a single citation in the literature are assigned to the bin "unknown." Then up march the categories: "less known" (1–9 citations), "known" (10–49), "well-known" (50–99), and so on. The highest category, "renowned," is reserved for those rare papers that have been cited at least 500 times.[10] By this measure, both the Brans-Dicke article and the Higgs articles became "renowned," accumulating more than 500 citations by 1981; to this day, each of these papers remains within the top 0.01 percent most-cited physics articles of all time.[11] (Several years ago, as my own most-cited physics article inched closer to 100 citations, administrators of the database inserted a new bin. Whereas papers with 100–499 citations had previously been labeled "famous," now the "famous" category would be reserved for papers with at least 250 citations. I like to think of my paper, which to date has received more than 200 citations, as "almost famous." Of course my wife, the psychologist, has no shortage of theories about all this.)

Each of these renowned papers proposed to explain the origin of mass by introducing a new field, φ, and accounting for its interactions with other types of matter. They were published around the same time, with lengthy articles appearing in the same journal, the *Physical Review*. And yet for twenty years, hardly any physicists considered the Brans-Dicke field and the Higgs field together. All told, 1,083 articles were published through 1981 that cited either the Brans-Dicke paper or the Higgs papers. Only 6 of these—less than 0.6 percent—cited *both* Brans-Dicke and Higgs, the earliest in 1972 and the rest after 1975.[12] The 1,083 articles in question were written by 990 authors. Only 21 authors cited both the Brans-Dicke article and Higgs's work—usually in separate papers—between 1961 and 1981. In other words, even as each set of ideas became "renowned" within its own subfield, *nobody* suggested that the Brans-Dicke field and the Higgs field might be physically similar, or even worth considering side by side, before the mid-1970s.

::::

The divide between particle physics and cosmology was especially sharp in the United States when Brans, Dicke, Goldstone, and Higgs introduced their respective φ fields. The Physics Survey Committee of the US National Academy of Sciences, for example, issued a policy report in 1966 entitled *Physics: Survey and Outlook*. The committee recommended that both funding and PhD-level personnel for American particle physics be doubled over the next few years—by far the largest increases suggested for any subfield of physics—while calling for virtually no expansion of the already-small areas of gravitation, cosmology, and astrophysics.[13] At a time when some of the most influential Soviet textbooks on gravity and cosmology began by discussing the latest speculations about nuclear forces, meanwhile, such a blurring of genres remained totally absent from American textbooks.[14]

The US research patterns—so starkly laid out in the 1960s policy reports and mirrored in the separate treatments of the Brans-Dicke and Higgs fields—were not set in stone. Indeed, by the late 1970s the separation between cosmology and particle physics no longer seemed quite so extreme; signs of a new subfield, particle cosmology, could be discerned, as the Brans-Dicke and Higgs papers began to chalk up more shared citations. Physicists have tended to account for the rapid rise of particle cosmology by appealing to new ideas from the mid-1970s, arguing that the power of these ideas alone compelled particle physicists to begin thinking about cosmology. Those new ideas included "asymptotic freedom," first published in 1973, and the construction of the first "grand unified theories," or GUTs, in 1973–74.

Asymptotic freedom refers to an unexpected phenomenon within certain models of forces that remain symmetric under local transformations: the strength of the force can decrease as the particles move closer together, rather than increasing the way most other forces do. For the first time, particle theorists were able to make accurate and reliable calculations of such phenomena as the strong nuclear force—the force that keeps quarks bound within protons and neutrons—as long as they restricted their calculations to very short distances. Such short distances corresponded to interactions at very high energies, far beyond anything that had been tested experimentally.[15] (H. David Politzer, David Gross, and Frank Wilczek shared the 2004 Nobel Prize for their discovery of asymptotic freedom.)

The introduction of GUTs also helped to point particle theorists' attention toward very high energies. As physicists were piecing together the Standard Model, several noticed that the strengths of each of the three forces described by the model—electromagnetism, the weak nuclear force, and the strong nuclear force—might become equal at very high energies. Theorists hypothesized that above that energy, the three forces would act as a single, undifferentiated force—hence the "grand unification." Below those energies, the GUT symmetry would be spontaneously broken, leaving three distinct symmetries describing three separate forces, each with its characteristic strength.[16]

The energy scale at which grand unification might set in was literally astronomical: more than one trillion times higher than anything particle physicists had been able to probe using Earth-bound particle accelerators. Physicists had no possible way of accessing such energies via their traditional route; even with four decades of improvements in the underlying technology, today's most powerful particle accelerators have increased the energies under study by a factor of a thousand, a far cry from a trillion. So GUT-scale energies could never be created in physicists' laboratories. But some began to realize that if the entire universe had begun in a hot big bang, then the average energy of particles in the universe would have been extraordinarily high at early times in cosmic history, cooling over time as the universe expanded. With the advent of asymptotic freedom and GUTs, particle physicists therefore had a "natural" reason to begin asking about the high-energy early universe: cosmology would provide what many came to call "the poor man's accelerator."[17]

Is this the whole story? Although certainly important, these new ideas, taken on their own, cannot explain why the new subfield, particle cosmology, emerged and grew as it did. For one thing, the timing is a bit off. Publications on cosmology (worldwide as well as in the United States) began a steep rise *before* 1973–74, and the rate of increase was completely unaffected by the appearance of the papers on asymptotic freedom and GUTs. Moreover, although GUTs were introduced in 1973–74, they did not receive much attention — even from particle theorists — until the late 1970s and early 1980s. Three of the earliest review articles on the emerging field of particle cosmology, published between 1978 and 1980, ignored asymptotic freedom and GUTs altogether, highlighting other work instead, some of it dating back to 1972, before either asymptotic freedom or GUTs had even been introduced.[18]

More than just ideas were at stake in the creation of particle cosmology. Politics, institutions, and infrastructure played major roles as well. When the Cold War bubble burst, right around 1970, academic physics fell into a tailspin. Nearly all fields of science and engineering entered a period of decline around that time; yet physics fell faster and deeper than any other field. Funding for physics fell about as quickly as enrollments did, plummeting by more than one-third between 1967 and 1976 (in constant dollars). By the early 1970s, physicists in the United States faced the worst crisis their discipline had ever seen.[19]

The cuts did not fall evenly across the discipline. Particle physics was hit hardest by far. Federal spending on particle physics fell by half between 1970 and 1974 (a combination of direct cutbacks and inflation), combined with a sudden drop in government demand for high-energy physicists.[20] The sudden cuts drove a rapid outflow of particle physicists: between 1968 and 1970, twice as many physicists left particle physics as entered it in the United States. The downward slide continued into the 1970s: the number of new particle physics PhDs trained per year in the United States fell by 44 percent between 1969 and 1975 — the fastest decline of any subfield. As particle physicists' fortunes tumbled, meanwhile, astrophysics and gravitation became some of the fastest-growing subfields in American physics. Spurred in part by a series of new discoveries during the mid-1960s (such as quasars, pulsars, and the cosmic microwave background radiation), as well as by innovations in experimental design, the number of new PhDs in this area per year grew by 60 percent between 1968 and 1970 and by another 33 percent between 1971 and 1976—even as the total number of physics PhDs fell sharply.[21]

Surveying the wreckage a few years into the slump, the Physics Survey Committee released a new report, *Physics in Perspective* (1972). The committee noted that theoretical particle physicists had fared worst of all when the cutbacks hit. When demand for particle physicists fell off, too many of the young particle theorists had difficulty switching their research efforts elsewhere. The nation's physics departments needed to revamp how particle theorists were trained, urged the elite committee:

> The employment problem for theoretical particle physicists appears to be even more serious than it is for other physicists. The large number of such theorists produced in recent years and their high degree of specialization are often given as the causes of this difficulty. This narrow specialization is already an indication that the student of particle theory has been allowed to choose unwisely, because real success in any part of physics requires more breadth. . . . University groups have a responsibility to expose their most brilliant and able students to the opportunities in all subfields of physics.[22]

Particle theorists were the only subfield singled out for such criticism in the entire 2,500-page report. Curricular changes quickly followed, aimed to broaden graduate students' exposure to other areas of physics, including more emphasis on gravitation and cosmology. Across the country, physics departments began to offer new courses on the subject. American publishers pumped out scores of new textbooks on gravitation and cosmology—having all but ignored the topic for decades—to meet the sudden demand. Whereas a major textbook publisher had advised series editors to proceed with caution when considering textbooks in the tiny field back in 1959—"There is probably not a vast market for a [general] relativity book, however good," one noted—publishers in the United States brought out twenty-six new graduate-level textbooks on the subject during the 1970s. Amid the fast-changing curricula, physicists sometimes decided not to wait for formal textbooks to be published. In 1971, for example, Caltech began to circulate mimeographed copies of the lecture notes from Richard Feynman's 1962–63 course on gravitation, while a Boston-based publisher rushed out another physicist's informal lecture notes on general relativity in 1974.[23]

These massive changes in American physics left their mark on the way theorists handled such esoteric ideas as the Brans-Dicke and Higgs fields. In 1979, two physicists in the United States independently suggested that the Brans-Dicke and Higgs fields might be one and the same—this after two decades in which virtually no one had even mentioned the two fields in the same paper, let alone considered them to be physically similar. Anthony Zee and Lee Smolin separately introduced a "broken-symmetric theory of gravity" by combining the Brans-Dicke gravitational equations with a Goldstone-Higgs symmetry-breaking potential, in effect gluing the two aspects of the φ fields together.[24] (Similar ideas had been broached tentatively by theorists in Tokyo, Kiev, Brussels, and Bern between 1974 and 1978, though they received very little attention at the time.)[25] In this model not only could the local strength of gravity, governed by Newton's "constant," $G \sim 1/\varphi^2$, vary over space and time (as in the Brans-Dicke work), but its present-day value emerged only after φ settled into a minimum of its symmetry-breaking potential (as in the Goldstone-Higgs work). In this way, Zee and Smolin aimed to explain why the gravitational force is so weak compared with other forces: when the field settles into its final state, $\varphi = \pm v$, it anchors φ to some large, nonzero value, pushing $G \sim 1/v^2$ to a small value.[26]

Anthony Zee's path to uniting the two φ fields illustrates one way in which physicists in the United States wandered into cosmology from particle theory after the collapse of the Cold War bubble. He had worked with gravitation expert John Wheeler as an undergraduate at Princeton in the mid-1960s before pursuing his PhD in particle theory at Harvard, earning his degree in 1970 just as the biggest declines in that area began. As he later recalled, cosmology had never even been mentioned while he was in graduate school. After postdoctoral work, Zee began teaching at Princeton. He swapped apartments with a French physicist while on sabbatical in Paris in 1974, and in his borrowed quarters he stumbled upon a stack of papers by European physicists who tried to use ideas from particle physics to explain various cosmological features, such as why our universe contains more matter than antimatter. Although he found the particular ideas in the papers unconvincing, the chance encounter reignited Zee's earlier interest in gravitation. Returning from his sabbatical, and back in touch with Wheeler, Zee began to redirect his research interests more and more toward particle cosmology.[27]

Lee Smolin, on the other hand, entered graduate school at Harvard in 1975, just as the curricular changes began to take effect. Unlike Zee, Smolin took courses in gravitation and cosmology alongside his coursework in particle theory—he didn't need to stumble into one area from the other. Smolin worked closely with Stanley Deser (based at nearby Brandeis University), who was visiting Harvard's department at the time. Deser was one of the few American physicists who had taken an interest in quantum gravity by the 1960s—attempting to formulate a description of gravitation that would be compatible with quantum mechanics. He was also the very first physicist in the entire world to publish an article that cited both the Brans-Dicke work and the Higgs work (although he treated the two fields rather differently and in separate parts of his 1972 paper). Smolin's other main adviser was Sidney Coleman, a particle theorist who just a few years earlier had begun teaching the first course on general relativity to be offered in Harvard's physics department for nearly twenty years. Smolin completed coursework with Steven Weinberg, whose influential textbook, *Gravitation and Cosmology* (1972), had recently appeared. Meanwhile, Smolin also took intense courses on Standard Model physics and GUTs with several architects of the new material, including Howard Georgi and visiting professor Gerard 't Hooft. Building on this broader range of courses, Smolin focused on quantum gravity for his research and suggested that the Brans-Dicke and Higgs fields might be the same just as he was finishing his dissertation in 1979.[28]

Smolin's experiences marked the new routine for his generation of theorists, trained during the mid- and late 1970s to work at the interface of gravitation and particle theory. Theorists like Paul Steinhardt, Michael Turner, Edward "Rocky" Kolb, and others—each of whom, like Smolin, received his PhD between 1978 and 1979—devoted formal study to gravitation as well as to particle theory in graduate school. Soon Smolin, Steinhardt, Turner, Kolb, and others were training their own graduate students to work in the new hybrid area. For these young theorists and their growing numbers of students, it became "natural" to associate the Brans-Dicke and Higgs fields with each other. Turner, Kolb, and Steinhardt each led groups that pursued further links between the two φ fields during the 1980s, constructing cosmological models in which the Brans-Dicke and Higgs fields either appeared side by side or were identified as one and the same. Some became avid program builders as well. Kolb and Turner, for example, established the first Center for Particle Astrophysics in 1983, carving out space for the new types of studies within Fermilab. They went on to write the first textbook for the new subfield, *The Early Universe*, which appeared in 1990.[29]

:::

I was a sophomore in college when Kolb and Turner's *The Early Universe* came out. Thanks to books like theirs, students like me could begin taking courses in particle cosmology as undergraduates. I was hooked immediately, thanks in no small measure to inspiring teachers—younger physicists who had studied in places like Fermilab's Center for Particle Astrophysics during their own training—and to Kolb and Turner's book. (I eventually purchased three copies, so that at least one would be near my fingertips at any moment.) For students of my own generation, it became routine, even second nature, to draw upon theoretical objects like the Brans-Dicke and Higgs fields in our research, hardly thinking twice about a move that had been so novel a few decades earlier. Indeed, from today's vantage point, it seems downright bizarre that physicists never considered the Higgs and Brans-Dicke fields in the light of one another for so long. The fields' union moved from unthinkable to unnoticeable within a few academic generations.

That seeming naturalness—the banality of combining those fields today and the strangeness of holding them at arm's length—illustrates how the contours of intellectual life can be reshaped by rapid changes in institutions and infrastructure, ultimately shifting the boundaries of what young physicists come to find compelling or worth pursuing. Hence the immediate appreciation I felt—tinged with a touch of sadness—when I stumbled upon the "Higgs inflation" paper while browsing the arXiv preprint server that morning, back in the autumn of 2007. Why didn't I think of that?[30]

COSMOS

14

Guess Who's Coming to Dinner

My mother rarely calls to talk about my research. One of the few exceptions occurred in April 2010, when she asked, "Do you agree with Stephen Hawking?" That's usually an easy question to field. On topics ranging from the behavior of black holes to the structure of the early universe, a safe answer is usually "Yes." But that wasn't what my mother was asking about. She was eager to learn whether I agreed with Hawking that trying to contact aliens would be a bad idea. Any extraterrestrial civilization that could receive our communiqués and act on them, Hawking warned, might show up on our doorstep looking to stay—and not in that friendly houseguest sort of way. "Such advanced aliens," Hawking surmised, might be "looking to conquer and colonize whatever planets they can reach."[1] In no time at all, the word spread from Hawking's voice synthesizer to the world's blogosphere. Soon even my mother was calling.

And so it was that the word "aliens" seemed to be on

everyone's lips and screens that spring, just in time to mark the fiftieth anniversary of SETI, the Search for Extraterrestrial Intelligence. Though philosophers and poets have long dreamed about alien intelligences, the modern history of SETI began with a brief article in the leading scientific journal, *Nature*. Two astrophysicists at Cornell University, Giuseppe Cocconi and Philip Morrison, postulated in 1959 that there might exist one uniquely well-suited frequency, nestled in the microwave region of the electromagnetic spectrum, at which intelligent civilizations might seek to communicate with us. Frank Drake, an astronomer at a newly established radio astronomy observatory in West Virginia, reasoned along similar lines. In 1960, he conducted his own search of the skies, under the code name "Project Ozma," hoping to catch some telltale sign of intelligence chiming in at the special frequency. He was greeted mostly with hiss; one heart-thumping squawk, he later determined, came not from the sky but from a top-secret military installation nearby. Not easily discouraged, Drake attracted other colleagues to the topic, and the search for ET began.[2]

Cocconi's and Morrison's *Nature* article makes for fascinating reading today. The article appeared less than two years after *Sputnik* had been launched, and it combined hardheaded calculation with an almost giddy optimism, the "can-do" and "gee-whiz" spirit that often marked the early years of the space age. Why look for signals from aliens? Because there are so many stars out there, Cocconi and Morrison explained. Many are similar to our Sun. Thus, Earthlike conditions, in which our own species evolved, might be fairly common throughout the galaxy. Cocconi and Morrison were further convinced that those myriad civilizations had likely developed "scientific interests" and "technical possibilities much greater than those now available to us."

Where there are Earth-like conditions, they reasoned, there could be life. Where there was life, there would be science.[3]

Throughout their brief paper, Cocconi and Morrison performed a strange rhetorical loop-the-loop. Given the state of our science and technology, how should we anticipate that advanced aliens would try to contact us? Humans had recently learned about a process in hydrogen atoms that emitted microwaves at a particular frequency: the so-called "21-centimeter line," first measured in a Harvard laboratory in 1951. Since hydrogen was the simplest and most abundant element in the universe, surely it provided a "unique, objective standard of frequency, which must be known to every observer in the universe," Cocconi and Morrison wrote; after all, *we* already knew about it. According to their calculation, the frequency of that special line lay in a sweet spot of the electromagnetic spectrum, away from naturally occurring sources of background noise. Given the universal nature of the special frequency, aliens might reasonably expect any civilization to design sensitive receivers tuned to that frequency early in their development of radio astronomy, as indeed we had done. Therefore, the only "rational" choice the aliens could make would be to broadcast their messages at that frequency, confident that some day we would follow the same scientific-technological developmental pathway that they did—or that we imagined they did, given our own recent experiences. Reasoning about others was inevitably a projection of ourselves.[4]

One needn't be a psychoanalyst or hold a PhD in cultural studies to discern a sweet, quiet yearning throughout Cocconi and Morrison's article. Not only were advanced extraterrestrial civilizations likely to exist, they asserted. But also the aliens were probably gentle, benign elders, monitoring our stellar neighborhood "expecting the devel-

Figure 14.1. *Above*, Astrophysicist Giuseppe Cocconi gives a lecture at CERN, 1967. (*Source*: Courtesy of CERN.) *Left*, Astrophysicist Philip Morrison. (*Source*: AIP Emilio Segrè Visual Archives, *Physics Today* Collection.)

opment of science near the Sun" and "patiently" beaming out their signals to us, ever hopeful that our return beacon might announce that "a new society has entered the community of intelligence."[5] Before Morrison began his work on SETI, he had served on the wartime Manhattan Project to design nuclear weapons. He inspected both Hiroshima and Nagasaki just weeks after the bombings in 1945 as part of the first scientific survey team. Seared by the experience, Morrison turned his energies to the budding arms-control movement. During the early 1950s, he was hounded by red-baiting critics for what they considered radical "world government" ideas.[6] No wonder he turned to the skies soon after in search of a more rational, welcoming community of civilizations.

SETI thus emerged as a by-product of the nuclear age. Drake, the radio astronomer, picked up where Cocconi and Morrison left off. To help organize discussions for a small workshop on SETI in 1961, he jotted down an equation, now known as the "Drake equation." He wanted some means of estimating how likely it might be that advanced alien civilizations were out there. Variables included the average rate at which new stars form; the fraction of those stars that form planets; the fraction of those planets that develop conditions suitable for life; and so on. The final term in his equation, L, denoted the average lifetime of alien civilizations.[7] Where Cocconi and Morrison's paper reflected the hopeful buoyancy of the early space age, Drake's equation bore the marks of its Cold War origin. For Drake as for nearly all his colleagues, L was always a stand-in for all-out nuclear war. Cocconi and Morrison assumed that life led inexorably to science. Drake continued: science led inexorably to nukes.

Paul Davies tackles the air of inevitability that animated the SETI pioneers in his book *The Eerie Silence: Renewing Our*

Figure 14.2. Frank Drake (*standing, second from right*) and other members of the original Project Ozma team gathered at the National Radio Astronomy Observatory in Green Bank, West Virginia, for a reunion in 1985, shown here posing in front of the Howard E. Tatel eighty-five-foot radio telescope used in the original 1960 Ozma experiments. (*Source:* Courtesy of NRAO/AUI/NSF.)

Search for Alien Intelligence (2010). Trained as a theoretical physicist, Davies now works in cosmology and astrobiology and heads the Beyond Center for Fundamental Concepts in Science at Arizona State University. (Some of his most influential early work refined physicists' notion of the vacuum.

He literally knows everything about nothing.) His main caution in the book is not to confuse necessary and sufficient conditions. The presence of water or amino acids on some distant planet seems necessary for life (at least for life as we know it); but their mere presence is far from sufficient for life to emerge. Same with the existence of Earth-like planets orbiting Sun-like stars. At the time that Cocconi, Morrison, and Drake were formulating their search strategies, astronomers had no direct evidence of planets outside our solar system. Today astronomers have identified thousands of these "exoplanets," and improvements in observing techniques promise to reveal many more. Davies rightly points out, however, that even if exoplanets do turn out to be exponentially plentiful in our galaxy, the spark of life might prove to be an even more exponentially improbable occurrence. The easy leap made in the early days of SETI—from stars to planets to life to intelligent life—was never more than a conjecture. Hence, Davies concludes, the "eerie silence"—no confirmed SETI contacts despite fifty years of concerted listening—might simply mean that life as we know it is rare, not that civilizations inevitably self-destruct in nuclear holocaust.[8]

Davies revisits other assumptions behind the early SETI work. Where Cocconi and Morrison assumed that intelligent civilizations would inevitably pursue scientific investigations, Davies counters that science is not universal, even here on Earth. Moreover, the old idea, common in the 1950s and 1960s, that basic science leads inevitably to improved technology seems difficult to square with humans' own historical record. Ancient Chinese civilizations developed astounding technologies but little of what might look like Western-style science. If science and technology could follow such contingent paths just among members of our own species over a relatively short time period (cosmologically speaking), why should we assume that extraterrestrial civilizations would march, lockstep, from intelligence to science to technology?

Davies questions the universality of science in one sense: the pursuit of scientific inquiry, the activity of science. An even more radical question concerns the universality of our scientific knowledge. Is our scientific representation of the natural world universal? Cocconi, Morrison, Drake, and their followers throughout the half century of SETI have argued over which regions of the electromagnetic spectrum would be most "rational" to target for a search. They have based their arguments on naturally occurring processes like the 21-centimeter hydrogen line or similar emissions from water molecules. But who is to say that other advanced civilizations—even if they do pursue something like scientific investigation—would carve up the buzzing, blooming confusion of nature in the same way that we do? Is our knowledge path-independent? We now think in terms of atoms, electrons, quantum transitions, and electromagnetic waves. Are those the only ways of making sense of physical phenomena? Is the intellectual history of Western science a universal pathway, some fixed point in the evolution of intelligence everywhere in the cosmos?

The question of universality enters Davies's account in a different way. He has chaired the SETI Post-detection Taskgroup of the International Academy of Astronautics. The committee's charge is to develop a protocol to be followed in the event that someone detects signals of possible extraterrestrial origin. Few topics today elicit as much blog-addled conspiracy talk as purported government cover-ups of UFOs and alien contacts. Davies's group has therefore aimed to steer a middle path between military-styled secrecy and unfettered airing of every false alarm. The current protocol calls for credible evidence to be shared first with other astronomers by means of the "Central Bureau for Astronomical Telegrams of the International Astronomical Union." (One can't help but smile at the name: telegrams to announce the findings, indeed.) The international community of professionals could then vet the evidence and try to rule out possible alternative explanations, like the terrestrial signals that fooled Frank Drake back in 1960. Next, the discoverer of the putative signal should alert the International Telecommunications Union, the International Council of Scientific Unions, and finally the secretary-general of the United Nations. Only after those international associations have been informed should the discoverer announce the finding to the public.[9]

Conspicuous in their absence from the long list of "telegram" recipients are any national governments. In large part, that is because SETI activities no longer receive government funding. The US National Aeronautics and Space Administration (NASA) once had a hand in the SETI game, but not anymore. Philip Morrison, of the Cocconi-Morrison search strategy, received the first NASA grant for SETI research in 1975. Soon money was flowing to research groups across the country. With much fanfare, NASA inaugurated its own SETI observational program on Columbus Day in 1992, five hundred years after the explorer reached the New World. All told, NASA spent nearly $57 million on SETI between 1975 and 1993 and had pledged an additional $100 million—modest sums compared to most "big science" appropriations but real money nonetheless. But the US Congress killed all federal funding for SETI one year after the

1992 Columbus Day festivities. Ever since then, SETI activities in the United States and around the world have been supported by private donations.[10]

SETI's difficulties in Congress stemmed in part from a wave of budget-cutting during the fall of 1993. Indeed, the SETI debates proved to be a dress rehearsal for bigger targets. A few weeks after ending government funding for SETI, legislators killed funding for the Superconducting Supercollider. The Supercollider carried a price tag one thousand times greater than NASA's annual expenditures on SETI, but both were felled by the same axe. Unlike the Supercollider, SETI operations were small and efficient, well managed and on budget. But SETI had few contractors to rally to its defense and little political "pork" to distribute across congressional districts. SETI also fell through the cracks between disciplines. It used tools from physics and astronomy without being central to either field, even as it trod on the toes of biologists, who knew about life, evolution, and intelligence.

Beyond the budget cuts and the usual political horse trading, SETI suffered from an image problem. Advocates talked about the "giggle factor." Grandstanding politicians wondered why the government should spend millions of dollars to search for extraterrestrial intelligence when one could just plunk down "75 cents to buy a tabloid at the local supermarket," as one Congress member stormed in 1990. Upon introducing the final amendment to kill SETI funding in 1993, one senator proclaimed, "This hopefully will be the end of Martian hunting season at the taxpayer's expense." Silicon Valley entrepreneurs and other private philanthropists largely picked up the tab, enabling SETI researchers to keep the project running.[11]

With all its talk about aliens, SETI has often been lumped

together with occult topics or pseudosciences. Searching for references to Cocconi and Morrison's article on the internet today, one finds links to the paper alongside books by New Age shamanism expert Carlos Castaneda and mystico-environmentalist tracts on the Gaia hypothesis. In the absence of confirmed contacts, critics charge, the field has been sustained by faith, hope, and speculation. Not so fast, proponents counter. SETI received high marks from three decadal reviews by the US National Academy of Sciences. The drive to improve detection capabilities led to major advances in microwave electronics and signal extraction. Frank Drake's first SETI search relied on a single-channel receiver. By the mid-1990s, SETI devices could simultaneously scan 250 million channels with state-of-the-art resolution. No wonder the US Federal Aviation Administration and the National Security Agency both expressed interest in SETI spin-offs. Other SETI techniques were quietly incorporated into next-generation methods to simulate the inner workings of thermonuclear weapons.[12] SETI still cannot shake the nuclear specter.

Davies has high hopes for a different kind of spin-off. SETI can inspire greater interest in science and encourage young people to ask big questions: not just about the enormity of space and the evolution of the cosmos but about the human condition and everything that unites us as a species. Davies anticipated Hawking's dour view of aliens, too. Although SETI does not broadcast signals of its own—it passively monitors for signals from others—Davies endorses "METI," or Messaging to Extraterrestrial Intelligence. He sees no reason to think that aliens would have any of the anthropomorphic traits (jealousy, colonizing impulses) that Hawking imputes to them. The message Davies would send? "Keep well clear and defend yourself": we are a helplessly militaristic society and armed to the teeth with nuclear weapons.[13] Davies sees Hawking's aliens in the mirror, and they are us.

In fact, SETI might make its grandest contribution in the nuclear arena. Rather than pseudosciences or the occult, SETI seems most similar to recent efforts by the US Department of Energy. The overseers of the nation's nuclear stockpile have had to get creative over the past two decades, designing secure facilities to house vast stores of radioactive waste. Some of the most hazardous by-products of the nuclear age, including isotopes of plutonium, have half-lives that stretch hundreds of thousands of years. The acutely toxic waste will be with us for countless millennia. One challenge is to find nooks and crannies on Earth that might remain geologically stable over those kinds of timescales, into which the waste can be buried. A second challenge is to design symbol systems that might warn our own future descendants, three hundred thousand years from now, not to go digging in bespoiled areas. As Harvard historian Peter Galison has documented, the US nuclear agencies have sought the wisdom of diverse experts—linguists, anthropologists, sculptors—to imagine how we might plausibly communicate with terrestrial beings in the impossibly distant future.[14] After all, the Latin alphabet dates back a mere twenty-six hundred years; cuneiform, the oldest known form of human writing, stretches back only a few thousand years before that. Only blind hubris could lead us to imagine that our familiar modes of communication will be recognizable in the year 300,000 CE.

Alongside the linguists and artists, the nuclear bureaucrats have also enlisted experts in SETI. Struggling to communicate with our own future selves calls for the same kind of radical imagination that SETI requires. Both efforts crisscross the boundaries between disciplines. Both face the challenge of extracting meaningful information from senseless noise. Both require experts to project from what we know about our own civilization to facilitate communication with some distant other. They are mirror images, twin children of the nuclear age, each haunted by the dangling L in Frank Drake's famous equation.

15

Gaga for *Gravitation*

A remarkable publishing event occurred in September 1973: the release of a 1,279-page book, weighing more than six pounds, with the simple title *Gravitation*. Wags were quick to remark that the book was not just about gravitation but a significant source of it. The book acquired several nicknames, including "the phone book" (another reference to its girth) and "the big black book" (for its sleek, modern cover). Most common became "MTW," named for the authors' initials: Charles Misner, Kip Thorne, and John Wheeler.[1]

Gravitation focuses on the general theory of relativity, Albert Einstein's remarkable theory of gravity. Einstein completed the theory just over a century ago. He had toiled for nearly a decade up to that point, stumbling through a series of false starts and working at a frenzied pace. Throughout the month of November 1915, he delivered regular updates on the emerging theory to the Prussian Academy of Sciences—one presentation each Thursday, four weeks in a row—adjusting details between each presentation. By the end of that month, he had arrived at a form of his equations that physicists still use today. Elegant and crisp, they are brief enough to tweet. Einstein's major insight was that space and time were actors in the story of nature, not merely a fixed stage on which all other activity played out. Space and time, on Einstein's account, were as wobbly as a trampoline—they could bend and distend in response to the distribution of matter and energy. That warping, in turn, affected objects' motion, diverting them from the straight and narrow path.[2]

One year after the armistice that ended the First World War, a British team, led by Arthur Eddington, announced that they had confirmed one of Einstein's key predictions: that gravity could bend the path of starlight. The dramatic announcement propelled Einstein and his general theory to instant stardom. Yet interest in the theory waned over the 1930s. Einstein himself noted plaintively, in a preface for a colleague's textbook in 1942, "I believe that more time and effort might well be devoted to the systematic teaching of the theory of relativity than is usual at present at most universities."[3]

Years passed, but eventually some charismatic teachers began to heed Einstein's call. Among the first and most influential was John Wheeler, who began to offer Physics 570, a full-length course on general relativity, at Princeton University in the mid-1950s. He quickly attracted world-class graduate students to the subject, including Charles Misner and Kip Thorne. Fifteen years later, concerned that textbooks on general relativity had failed to keep up with modern developments, Misner, Thorne, and Wheeler teamed up to write *Gravitation*.[4] Upon publication, *Gravitation* joined several other new books about general relativity, including

Steven Weinberg's *Gravitation and Cosmology* (1972) and Stephen Hawking and George Ellis's *The Large Scale Structure of Space-Time* (1973).[5] Unlike those other books, however, MTW defied many people's expectations for a textbook. Some just didn't know what to make of it.

Misner, Thorne, and Wheeler clearly intended *Gravitation* to be a textbook, pitched at advanced physics students. Wheeler's notes from an early planning meeting with his coauthors made clear that they would write the book with "the committee planning graduate courses in U. of X" in mind. While certainly thinking in terms of a textbook, however, from the start they treated the project as an experiment in the genre. They developed a rather complicated structure for the book, dividing material into two tracks: a core of introductory material occupying less than a third of the book, surrounded by extensions, elaborations, and applications.[6] The two tracks were not sequential; many chapters were divided, section by section, into one track or the other. Even more novel was the extensive use of "boxes" for complementary material. The boxes were set off from the main text by heavy black lines, interrupting the flow of ordinary chapter exposition, often for several pages at a time. Some of the boxes resembled the "sidebars" that had long been a staple of science textbooks aimed at younger students and featured short biographies of famous physicists or brief descriptions of important experiments. But most of the boxes in *Gravitation* served a different purpose. According to Wheeler's notes, the boxes were meant to constitute "a third channel of pedagogy," beyond the two tracks. "They are distinguished from the main text by untidiness" and included "the kinds of things we would like to present in lecture hour to students who can be relied upon to learn tightly organized material and computational methods on

Figure 15.1. *Top left*, Charles Misner. (*Source*: AIP Emilio Segrè Visual Archives, *Physics Today* Collection). *Top right*, Kip Thorne in 1977. (*Source*: Courtesy of the Archives, California Institute of Technology). *Bottom*, John A. Wheeler in his office, late 1970s. (*Source*: Photograph by Frank Armstrong for The University of Texas at Austin, courtesy of AIP Emilio Segrè Visual Archives, Wheeler Collection.)

their own from a systematic text." Their pedagogical aspirations were clear: as each author drafted a section of the book, the coauthors would "*test* a write up by asking if a student could use it to lecture from."[7]

The authors devoted spectacular attention to the physical appearance and production of the book. Thorne traded detailed letters with the artists and layout designers at the original publisher, W. H. Freeman in San Francisco, going over everything from the thickness of lines setting off the box material to arrow styles and shadings to be adopted in the hundreds of illustrations. Early on, Thorne alerted an editor at Freeman that "several features of the manuscript will require special typesetting problems." Beyond the extensive figures, tables, and boxes, the authors anticipated the need for at least six distinct typefaces, perhaps as many as eight, to properly distinguish the plethora of symbols and equations they would be treating.[8] (Before the book had even been published, Thorne worried that "the extreme complexity of the typography" would bedevil foreign-language publishers. He recommended that they simply photograph the equations from the English edition once it became available rather than attempt to re-typeset them.)[9] Given the book's unusual organization, the authors also inserted thousands of marginal comments throughout the book. Some comments summarized the material under discussion, but many others were "dependency statements": a road map spelling out at each point in the massive tome which other sections a given discussion depended upon, and which others would in turn depend upon it.[10]

Having tackled every detail of composition and typesetting, imagine the authors' surprise when—two years into the process, and just three weeks before they submitted their final, edited manuscript—they learned that the publisher held a rather different conception of the book than they did. After meeting with their editor from the press, Thorne shot off a letter to his coauthors. "I was rather shocked to learn from Bruce [Armbruster, the editor] that the people at Freeman are so out-of-touch with our book that they have not been regarding it as a textbook, but rather as a technical monograph. I suppose that the enormous size of the book has something to do with it." The publisher's plan had been to produce an expensive hardcover edition, intended primarily for purchase by libraries: "Freeman had not been expecting to pick up the textbook market with this book" at all. Thorne worked hard to convince the editor that "there might be some hope of picking up student sales" as well, but that would require a complete overhaul of the publisher's printing and pricing plans.[11]

Was *Gravitation* a reference monograph for libraries or a textbook for classroom use? From that ontological difference sprang more immediate considerations. For example, how could they keep such a fabulous concoction from crumbling under its own weight? The book's unusual trim size — each of its nearly 1,300 pages was more than an inch wider and taller than standard textbooks at the time — suggested hardcover rather than paperback binding. Hardcover binding seemed all the more appropriate to the authors, for whom *Gravitation* was self-evidently a textbook, since (as Thorne explained) "it seems to me that paperback editions cannot hold up well enough with the heavy use that a student in a full year course would give the book." But hardcover binding threatened to price the book beyond the reach of a student market.[12] After assurances from the publisher that paperback binding could hold up just as ruggedly as hardcover, the authors struck a deal with the publisher: in exchange for reduced royalty rates on the paperback edition, the press would aim to keep the price of the paperback lower than the hardcover price of the recent textbook by Weinberg, *Gravitation and Cosmology*. Upon publication, the paperback edition of Misner, Thorne, and Wheeler's *Gravitation* sold for $19.95 (about $110 in 2020 dollars), and the hardcover for twice that price. With the publisher now treating the book as a textbook rather than a reference monograph, and with the compromise pricing plan in place, Thorne was confident that the book could "capture one hundred percent of the textbook market in this field—or as nearly so as possible."[13]

Like the authors and publisher, reviewers recognized the book as unusual. "A pedagogic masterpiece," announced a reviewer in *Science*; "one of the great books of science, a lamp to illuminate this Aladdin's cave of theoretical physics whose genie was Albert Einstein," crowed another in *Science Progress*. A third reviewer challenged his readers: "Imagine that three highly inventive people get together to invent a scientific book. Not just to write it, but invent the tone, the style, the methods of exposition, the format." Many reviewers lauded the rich set of illustrations and the innovative use of boxes.[14] Others complained that the two-track-plus-box organization introduced too many redundancies. "This is a difficult book to read in a linear, progressive fashion," concluded one reviewer; "there is needless repetition (indeed, almost everything is stated at least three times)," noted another. "The variety of gimmicks is bewildering— framed headings with quotations, marginal titles, 'boxes' sometimes extending over several pages, heavy type, light type, large type, small type," reported a reviewer in *Contemporary Physics*. "Clearly the book is an experiment in presentation on a grand scale."[15]

Nearly all reviewers commented on the writing style.

Wheeler was already well known among physicists for his catchy slogans and engaging prose. (Among other memorable contributions, Wheeler had coined the term "black hole.") Wheeler's early planning notes for the book insisted that he and his coauthors must "make clear the idea itself. But soberly, factually, no hyperbole, no enthusiasm."[16] If that had been the intention, not all reviewers agreed on the outcome. The book featured a "prose style varying from the unusually colloquial to the unusually lyrical," wrote one reviewer. But one person's lyricism was another's doggerel. "There is a commendable attempt at informality, but this reviewer found the breeziness irritating at times," came one verdict. "A 'poetical' style is understandable if one deals with such [speculative] topics as 'pregeometry.' However, 'poetical' passages in differential geometry, for example, may obstruct the understanding of an ascetic reader," concluded another.[17] One reviewer huffed that the informal writing style "comes dangerously close to being patronisingly simplistic, to the point of insulting the reader's intelligence." Another reviewer was even more scandalized by the book's tone. The intended reader, he scoffed, would be most at home with the book "if he is a regular subscriber to *Time* magazine—the writing of these authors has much in common with its breathless style."[18] Subrahmanyan Chandrasekhar, the Nobel-laureate astrophysicist who had grown up in India, trained in Britain, and settled in the United States, likewise noted that the book's "style fluctuates from precise mathematical rigor to evangelical rhetoric." He closed his review with the memorable observation: "There is one overriding impression this book leaves. 'It is written with the zeal of a missionary preaching to cannibals' (as J. E. Littlewood, in referring to another book, has said). But I (probably for historical reasons) have always been allergic to missionaries." (Thorne wrote to Chandrasekhar that the closing paragraph had left him "chuckling for about ten minutes.")[19]

While acknowledging the book's unusual organization, writing style, and pedagogical innovations, most reviewers treated the book as the authors had intended: as a textbook primarily for graduate-level coursework in the technical details of gravitational physics. The authors had set out to corner the market for textbooks on the topic, and they largely succeeded. A few years after publication, their book was still selling between four thousand and five thousand copies per year, while their main competitor, Weinberg's *Gravitation and Cosmology*, had dropped to around one thousand copies per year. Thorne noted to the publisher—with fanfare but not much hyperbole—that by the late 1970s, "a large fraction of the physics graduate students in the Western world bought a copy of *Gravitation*."[20] The book sold fifty thousand copies during its first decade, at a time when institutions in the United States graduated about one thousand PhDs in physics per year, and no other country came close to those annual totals.[21]

Yet from the start, some readers saw much more in *Gravitation* than a vehicle for training soon-to-be specialists. The publisher, for one, reversed course in a dramatic way. A decade after having written off the book as merely a reference work for library purchase, the editors decided to advertise a specially reduced price on the book—nearly 25 percent off list price—to subscribers of the popular magazine *Scientific American*. Thorne countered that a better way to test "the elasticity in the demand" for the book would be to offer that reduced price to "that portion of the market which concerns me most": students and young academics. He urged the publisher to offer the reduced price to university bookstores rather than *Scientific American* devotees.[22]

Nonetheless, the publisher was on to something. Upon the book's publication, reviews had run not just in venues such as *Science* and *Physics Today*; the *Washington Post* devoted a full-page review to the book, and a daily newspaper in San Antonio, Texas, likewise recommended it. The reviewer in the *Post*, himself a physicist at Williams College, acknowledged, "Perhaps it is strange to review here a textbook full of mathematics, a book, moreover, whose 6.7-pound bulk the young, the old and the infirm can scarcely lift. But," he declared, "those who read like to know what is being published and discussed." And *Gravitation* certainly warranted discussion. The book's engaging prose "awakens hope that the fuzzy and lugubrious 'style' that still spreads its gloom over so much of American science may not be in fashion forever." The book's unusual organization, moreover, seemed akin to recent trends in avant-garde filmmaking, such as the French *nouvelle vague*. "There are very few stories that should be told sequentially," the reviewer avowed. All the better that *Gravitation*, like the hip filmmakers, had discovered "strategies for breaking up a linear narrative."[23] The San Antonio reviewer likewise encouraged his readers. "I am not a mathematician, and the 200 or so pages I've read are not all that formidable," he explained. "If you're curious and have an imagination, you won't be cowed. The challenge is stiff, but fascinating." The organization of the book was "phenomenal," and the topic inspiring. He concluded, "This is a fabulous, rewarding book." Novelists could scarcely hope for a more enthusiastic review.[24]

Fan letters also streamed in to the authors from a wide assortment of readers. Many came from colleagues who reported how much they enjoyed teaching from the book in their formal classes.[25] But others came from further afield. One reader wrote from a hospital in Italy—it is not clear whether the handwritten letter came from a patient or a physician — to press the authors whether their views about the cosmos had changed during the three years since the book's publication. (The letter writer had been keeping up with more recent discussions in the field by reading the Italian-language edition of *Scientific American*.) He had more specific questions, too. In particular, what was the fate of life in a universe that cycled from big bang to big crunch? He was so desperate for a response that he promised $200 to anyone (the authors or their graduate students) who might take the time to answer. "Don't be offended by my proposal. Time = Money."[26]

An engineer in Brussels turned to the book for a different reason. He decided to pick up *Gravitation* to help him learn English before beginning military service. "My hopes have been completely fulfilled: *Gravitation* is worth reading to learn English because it makes enjoy Physics!" The book so inspired him that he drew seven full-page, whimsical cartoons in the style of Antoine de Saint-Exupéry's *The Little Prince* to illustrate concepts he had learned from *Gravitation*.[27]

Readers closer to home wrote to the authors as well. Especially poignant was a letter that Thorne received from a reader in Portland, Oregon. "I stumble here, fall down there, and generally make a fool of myself as I wander about your textbook," the correspondent explained, "but I am gaining a sense of balance and a few tools with which to deal with the subject." His dedication to the book was impressive:

> When friends ask me about what I am doing I have made
> the mistake of telling them the truth [about his attempts
> to read *Gravitation*]. Sometimes I think they are right,
> I feel as though I am on the brink of madness. I go out

to have a beer and listen to someone talk about his love affairs, the clutch in his pick-up truck, the problems with his children, the plumbing, the bus service. I look at him and see him dealing with all these important issues and I ask myself why do I care if I ever understand the difference between leptons and leprosy?

Yet still he could not shake his "obsession" with Einstein's own question, "whether or not God had any choice in the creation of the Universe." He needed to know: "Could God be a traveling technician whose responsibility is to supervise gravitational collapses and big bangs?"[28]

Six years after publication, with annual sales still brisk, John Wheeler tried to assess the reasons for the book's success. Writing to his editor, Wheeler surmised that "many people buy the book who are attracted by the mystique, the boxes, the interesting illustrations, the ideas but who don't expect to and never will get deep into the mathematics." He figured about half the purchasers fell into that category—and he was eager not to lose them. In thinking about revising and updating the book, Wheeler concluded that "I think we can add a few things and take away a lot of things to keep this group 'on board.'"[29] Those plans fell through—Misner, Thorne, and Wheeler never did undertake a revision of their massive masterpiece—but Wheeler's observation nonetheless rang true. In their effort to write a specialized textbook they had produced a hybrid work, as attractive to *Scientific American* subscribers for its "mystique" as to doctoral students struggling to enter the field.

Gravitation narrowly escaped the pigeonhole of library-only reference work and went on to sell tens of thousands of copies. The book received extensive analysis and review in physicists' specialist journals, even as it inspired passion—

even ecstasy—among journalists and nonspecialist readers. Somehow this hulking book, stuffed to overflowing with equations so complicated they required multiple typefaces and elaborate marginal notes, exerted broad, crossover appeal. For generations, the book—like Einstein's elegant theory at its core—has inspired alluringly large questions. Why are we here? What is our place in the cosmos?

I still cherish my own copy of the original edition of MTW, which I picked up at a used-book store during graduate school. By now the binding has all but disintegrated, the pages are a bit yellowed, and whole sections threaten to fall free. Yet I can't bring myself to toss it, even though Princeton University Press recently published a lovely, sturdy reprint edition.[30] Today the two editions sit side by side on my bookshelf, together hogging more than six inches of prime shelf space. When I want to double-check how those world-class teachers broached a particularly subtle subject, I grab the new edition. But when I want to feel—or at least imagine—some more direct connection with the generations of students and seekers who came before me, I reach for the old, with both hands.

16

The Other Evolution Wars

Will tomorrow be just like today? What about ten million tomorrows from now, or a billion yesterdays ago? Has the universe always been more or less as we observe it now, or is it cosmic destiny—as it is for people—to change and evolve over time? The question is hardly new. Plato reported that Heraclitus, before him, had observed that "all things pass and nothing stays," a stance conveyed in the gnomic (and oddly tweetable) phrase "you cannot step twice into the same river."[1]

Einstein, too, wondered about the nature of cosmic change. He focused on the question with fresh intensity soon after arriving at his general theory of relativity. With his new equations in hand, he began to wonder how his key insight about gravitation as the warping of space and time might play out not just for this or that local phenomenon—planets whirling around the Sun, the path of light rays bending as they approach a large mass—but for an entire universe at large. By design, his first models were simple, assuming, for example, that all the mass and energy within an imaginary universe were spread evenly throughout space. By 1917 he had launched a new scientific specialty, soon dubbed "relativistic cosmology": an attempt to describe whole universes within the rubric of general relativity.[2]

Right away, Einstein learned—not always happily—that other colleagues held strong ideas of their own. Several put Einstein's equations to work in ways that he found not just wrongheaded but repugnant; their cosmic visions seemed so much at odds with his own. Since that time, as younger generations have sought to sharpen and refine the earlier ideas, they, too, have found themselves sparring with each other, their debates shot through with a strange mix of mathematics, aesthetics, and religion. In our own time just as in Einstein's, questions about cosmic evolution remain entrenched within still harder questions about how we can know—about the universe and our place within it—and who gets to say.

::::

One of the first scientists to seek insights into how general relativity might describe the universe as a whole was a young Russian mathematical physicist, Alexander Friedmann. Friedmann was a native of Saint Petersburg. When he first encountered Einstein's relativity, soon after the end of the First World War, his city had been renamed Petrograd, in honor of the eighteenth-century tsar Peter the Great; by the time Friedmann produced his most ambitious cosmological work, early in 1924, the Bolsheviks had rechristened the city Leningrad. Amid such tumult, perhaps it is not surprising that Friedmann's forays into relativistic cosmology emphasized change over time, even anticipating violent disruptions. In a series of succinct papers, Friedmann demonstrated that Einstein's equations could describe universes that evolved, expanding from a minute speck to supergalactic scales. In some cases, Friedmann showed, a universe could halt its expansion and collapse back on itself. The determining factor—the quantity that would fix these cosmic destinies—was the average amount of matter and energy per volume. Overdense universes would bloat but then break; underdense universes would continue expanding forever.[3]

Einstein liked these options not one bit. He loathed the idea of cosmic expansion, or indeed of any large-scale evolution. Such changes over time lacked a certain aesthetic satisfaction, Einstein complained; he preferred the pristine symmetry of a universe that always was and always shall be. Even as Einstein was dismissing Friedmann's solutions, however, a young Belgian mathematical physicist, Georges Lemaître, was busy reproducing them. In addition to the allure of Einstein's equations, Lemaître was inspired by recent astronomical observations spearheaded by the American astronomer Edwin Hubble. Working in southern California with some of the world's largest telescopes, Hubble and his colleagues had found an intriguing relationship between how far away objects seemed to be from Earth and how quickly they were moving still farther away. Lemaître drew the conclusion even before Hubble did: our universe seemed to be expanding, distant objects moving ever farther apart from other objects over time. In 1931, Lemaître pressed further, arguing that if the universe is expanding today, then it must have been smaller in the past. Some finite time ago, all the matter in the universe must have been concentrated at

Figure 16.1. Albert Einstein (*right*) meets with Georges Lemaître (*center*) at the California Institute of Technology in the 1930s, joined by Caltech president Robert A. Millikan (*left*). (*Source*: Bettmann, courtesy of Getty Images.)

a single point. The universe, Lemaître concluded, must have begun in a very hot, dense state—he called it the "primeval atom"—and has been expanding ever since.[4]

Alongside his studies of physics and mathematics, Lemaître had pursued his other great passion: theology. He was ordained as a Catholic priest in 1923, and at least in these early days, his cosmology and theology seemed well integrated. In an early draft of his article on the primeval atom, he rhapsodized, "I think that everyone who believes in a supreme being" would be "glad" to see such congruence between science and religion. He struck out this passage just prior to publication—perhaps recognizing that articles in *Nature* rarely invoked God—and thereafter argued strongly against mixing theology and cosmology. He was an especially outspoken critic of biblical literalism, returning time and again to a position that Galileo had spelled out in his famous letter to Grand Duchess Christina back in 1615: the Bible teaches us how to go to Heaven, not how the heavens go.[5]

Lemaître's newfound care notwithstanding, several of his colleagues continued to advise caution—not just about mixing science and religion but about believing in a universe that had a beginning and has been evolving ever since. To some, such a scenario smacked too much of the Genesis account. Arthur Eddington—devoted Quaker, giant of British astrophysics, and onetime teacher of Lemaître's—responded that a universe that had a beginning in time might be physically possible but "philosophically it is repugnant to me." Richard Tolman, an accomplished physical chemist at the California Institute of Technology and one of the earliest champions of relativistic cosmology within the United States, went further still. He cautioned his colleagues in 1934, "We must be especially careful to keep our judgments unaffected by the demands of theology and unswerved by human hopes and fears."[6]

Many of these early cosmologists became best-selling authors. In their popular books, they freely debated each other's conclusions—scientific, aesthetic, religious, and otherwise. Yet their discussions received little pushback from nonscientists at the time. Just when "evolution" of a different sort was on many people's minds—with the Scopes trial of 1925 inspiring sensational coverage of Darwin's theory of biological evolution and forcing hard questions about whether such ideas could be taught in Tennessee's public schools—the cosmologists' evolutionary musings inspired more laughter than fear or anger.[7] Consider these snippets of advice from the *New York Times*: "Einsteinism: just ignore it as of no concern to us" (1923); readers should file modern physics under "things you needn't worry about just yet" (1928); modern cosmologists were just as quaint as medieval theologians counting the number of angels who can sit on the head of a pin (1931); and modern physics failed to answer life's most important questions (1939). While Einstein and his colleagues battled over the idea of an evolving cosmos, few outside their circle felt compelled to weigh in.[8]

:::

Soon after the Second World War, a trio of physicists working in the United States returned to Lemaître's ideas, now armed with new ideas and data about nuclear physics—data obtained from the wartime Manhattan Project, from the postwar efforts to design hydrogen bombs, and from ongoing work with nuclear reactors. The moving force behind the new work was Russian émigré George Gamow, who had first learned Einstein's general relativity from Alexander Friedmann in the 1920s. Joining Gamow were Robert Herman and Ralph Alpher, two young physicists at the Johns Hopkins Applied Physics Laboratory. The laboratory had been established in 1942 to work on military projects, such as the proximity fuze. Most of the projects after the war were missile projects for the Navy.[9] Immersed in details about defense systems, Alpher and Herman managed to reserve a portion of their time to think with Gamow about the cosmic and otherworldly. In particular, they worked hard to flesh out Lemaître's picture of a universe beginning in time and evolving through various stages. Their main question: where did the elements come from?[10]

Figure 16.2. In this composite image, George Gamow's face emerges from a bottle of "Ylem," an ancient Greek word meaning "substance." Gamow and his colleagues Robert Herman (*left*) and Ralph Alpher (*right*) used the word to refer to the primordial mixture of protons and neutrons from which heavier nuclei might have formed soon after the big bang. (*Source*: Public domain.)

Their answer, which formed the basis of Alpher's dissertation under Gamow's direction in 1948, came to be known as "nucleosynthesis." Gamow and company calculated that in the earliest moments after the beginning of the universe, ambient temperatures would have been unimaginably high. Energetic photons (individual particles of light) from the hot surroundings would each carry so much energy that they would blast apart nuclear particles, such as neutrons and protons, when those particles began to stick together. That is, the photons' energy would overwhelm the binding energy of the strong nuclear force, which would otherwise make the nuclear particles form atomic nuclei. As the universe expanded, however, it would cool, just like the gas inside an expanding balloon. At a calculable moment—roughly one second after the beginning—the force of nuclear attraction would begin to win out over the less-energetic photons, and neutrons and protons would begin to stick together in stable deuterium nuclei. As the universe continued to expand and cool, additional nuclear particles would glom on to these light nuclei, building up the heavier elements.

Never one to be shy about his findings, Gamow trumpeted his group's work in playful terms. He wrote to Einstein about this new account of "the Days of Creation" and titled one of his popular books *The Creation of the Universe* (1952), echoing the biblical term. Late in 1951, Pope Pius XII delivered a lecture before the Pontifical Academy of Sciences. Impressed by the easy fit between Gamow's developing model and scriptural accounts, the pope declared that the physicists' work "invokes no new ideas even for the simplest of the faithful. It introduces nothing different from the opening words of Genesis, '*In the beginning* God created heaven and earth. . . .'" To Gamow—an inveterate jokester—the pope's offering seemed too good to be true. Three months later he submitted a brief article to the *Physical Review* in which he quoted extensively from the pope's lecture, citing it as an authority for his latest research.[11]

Fred Hoyle, a British astrophysicist based at Cambridge University, did not find Gamow's pranks amusing. Hoyle had learned his general relativity during the 1930s from Arthur Eddington, and he, too, sought to build a coherent cosmological model soon after the war. Together with the Austrian transplants Hermann Bondi and Thomas Gold— each of whom had left the Continent to study at Cambridge before the Nazis overran Austria—Hoyle developed a rival cosmology to Gamow's. Hoyle, Bondi, and Gold argued that all astronomical observations to date could be accounted for by a steady-state universe. In their model, the universe had no beginning in time, and it has always been expanding. That second part—an eternal, nonstop expansion—could be possible, they reasoned, if a trace amount of new matter was constantly being created throughout space, far less than could be detected experimentally. If a tiny little bit of new stuff was constantly being made, then on average the universe would look the same at any given moment; there would be no evolution. In place of early-universe nucleosynthesis, Hoyle and company hypothesized that all atomic nuclei were cooked inside stars and then dispersed throughout the cosmos when heavy stars blew up in supernova explosions.[12]

More than physics seemed to be at stake. Hoyle spoke out vigorously against any theological incursions into physics. In his 1950 popular book, *The Nature of the Universe*, based on a series of radio lectures for the British Broadcasting Company, Hoyle charged that the very notion of a universe beginning in time was "quite characteristic of the outlook of primitive peoples," who turn to gods to explain physical phenomena. Ironically, Hoyle himself coined the term "big bang" to describe Gamow's program during these radio lectures; it was meant to sound childish and dismissive. Moreover, Hoyle and his colleagues insisted, physicists had little warrant to trust physical laws like general relativity and the behavior of nuclear forces when extrapolated so far from the regimes in which they had been tested. Big-bang advocates who stuck doggedly to such calculations, Hoyle charged, behaved just like Catholics and Communists: both, he said, were blind believers, too easily swayed by dogma.[13]

Although most physicists ignored Gamow's and Hoyle's colorful exchanges at the time, popular news media did cover the debate. Where once the *New York Times* had playfully chided physicists and cosmologists for their utter irrelevance, few drew the same conclusions after the Second World War. In the wake of wartime projects like radar and the atomic bomb, physics filled a special—indeed unprecedented—cultural niche, especially in the United States. In fact, the overwhelming majority of times that the phrase "big bang" appeared in major newspapers during the 1940s and 1950s, it referred not to Gamow's cosmology but to nuclear weapons testing and Cold War brinksmanship with the Soviet Union.

Perhaps the intertwining of cosmology, nuclear physics, and geopolitics explains another curious reaction to the postwar research. Though the postwar decades saw a resurgence in the United States of strident opposition to Darwinian evolution by evangelical Christians, few but the most hard-core biblical literalists rose to challenge the physicists' ideas about the big bang. Even the most influential advocates of "creation science" after the war drew distinctions between the age of the Earth—which they took to be roughly six thousand years old, as required, in their view, by the book of Genesis—and the age of the universe at large. John Whitcomb and Henry Morris's runaway best seller *The Genesis Flood* (1961), for example, argued that the biblical account of creation applied to the solar system but not to the entire cosmos.[14] Even as they argued passionately against standard geology and biology, these prominent creationists issued physics and cosmology a free pass. In the nuclear age, physics seemed untouchable.

: : :

Scientific consensus in favor of the big bang coalesced during the mid-1960s. Several new astrophysical observations lent strong support to the model but seemed irreconcilable with a steady-state universe.[15] Yet few in the field had time to take a victory lap. Before long, a whole new set of challenges emerged. A small group of physicists began to wonder if Einstein's framework for all these cosmological discussions—general relativity—was itself only an approximation. What if the universe was built, not out of point-like particles dancing on a curved spacetime, but out of tiny, one-dimensional "strings"? In that case, general relativity would be no more fundamental than Newton's physics: each would be a useful approximation, accurate for certain types of applications but not truly a law of nature. And if *that* was the case, then physicists would need to rethink relativistic cosmology.

String theory remained a marginal curiosity in the field for several years before bursting onto center stage in the mid-1980s. In a swift series of developments, later dubbed the "first string revolution," two groups of mathematical physicists demonstrated that string theory might offer a way to make gravity compatible with quantum theory, while avoiding many of the traps that had marred earlier efforts to combine the two. Such unification had long been the holy grail of theoretical physics: only a quantum theory of gravity could possibly be merged with the other known forces of nature, each of which is clearly quantum-mechanical in origin. String theory's champions began to proclaim it a "theory of everything."[16]

Today, string theory is undeniably at the forefront of high-energy physics. Since the early 2000s, physicists have routinely published two thousand articles or more on the topic each year. Yet as physicist Lee Smolin has argued, string theory is something of a "package deal": it combines many features that physicists want with others that are far less desirable. (Smolin compares the situation to the challenge of buying a new car: the sunroof you like comes packaged with a fancy stereo system you don't really need.) For starters, string theory requires an as-yet-undetected symmetry among the known particles, known as "supersymmetry." More troubling, the theory can only be formulated in ten spacetime dimensions, rather than the four in which we seem to live: one dimension of time, plus the three spatial dimensions of length, breadth, and height. Worst of all, at least according to some critics, string theory leads to a huge number of possible universes, with (as yet) no way to choose between them. It has gone, in effect, from a "theory of everything" to a "theory of anything."[17]

Since the 1980s, physicists have recognized that they need more than just knowledge of the governing laws of the theory in order to make specific predictions about how our universe should behave. They also need to know how the extra dimensions are arranged: are they curled up like tiny soda straws or twisted in some more complicated shapes? Every quantitative prediction from string theory depends on the (unknown) topology of the extra dimensions. For two decades, physicists thought the number of topologically distinct possibilities numbered in the hundreds of thousands.[18] The situation became exacerbated in 2000 when leading string theorists Joseph Polchinski (at the University of California at Santa Barbara) and Raphael Bousso (then a postdoctoral fellow at Stanford, now a professor at Berkeley) recognized that other structures—dubbed "fluxes" and "membranes"—could wrap around these extra dimensions. Instead of 10^5 possibilities, there appear to be upward of 10^{500} distinct low-energy states in string theory, any one of which (or none) might describe our observable universe. Every single observable quantity in our universe, from the masses of elementary particles, to the strengths of the fundamental forces, to the expansion rate of our universe, and more, would depend on precisely which of these stringy states our universe happened to be in. And yet string theorists to date have found no way to explain why our universe is in any particular one of these many possibilities.[19]

Pause for a moment to consider that number: 10^{500}. It is utterly removed from our everyday experience, all out of proportion to other numbers that scientists usually encounter. In fact, it is difficult to generate a number that large using familiar quantities. Let us start with the earthly and mundane: the ratio of billionaire Jeff Bezos's personal wealth (if internet accounts are to be believed) to my own is a measly 10^5—and whether that number seems encouraging or depressing, it is nowhere near 10^{500}.[20] Cosmic numbers likewise fail to come close. The age of our observable universe is about 10^{17} seconds; the ratio of the mass of the Milky Way galaxy to the mass of a single electron is roughly 10^{71}.

The story gets still more bizarre. Building on the now-standard supplement to the big-bang model—inflationary cosmology, which posits a brief burst of exponential expansion early in our universe's history—some string theorists began to argue that these 10^{500} states are not just theoretical possibilities but are actually out there, real "island universes" all their own. Central to the argument is that once inflation begins somewhere, it will continue forever. (This has been dubbed "eternal inflation.") In any given region of spacetime, the exponential expansion will halt after a characteristic period of time, much like the half-life of radioactive substances here on Earth. But in most inflationary models, this half-life is longer than the time it takes for a volume of space to double in size. So the volume of space that is inflating will always win out over the pockets that happen to stop inflating. On this view, we live within one "pocket" or "island" universe within a much larger "multiverse."[21] As theorists like Stanford University's Leonard Susskind and Tufts University's Alex Vilenkin see it, by combining the "landscape" of string possibilities with eternal inflation, the relevant question becomes not why one unique state got picked out of the huge number, but why *we* happen to live in the particular island universe that we do.

To tackle this question, Susskind, Vilenkin, and growing numbers of their colleagues have turned to something called the "anthropic principle." The natural constants in our observable universe—all those particle masses, force strengths, expansion rates, and so on that depend on which string state our universe occupies—must fall within rather narrow ranges in order for life as we know it to exist. Presumably, these constants would not be conducive to life (at least not life like us) in the vast majority of the other string states and hence in the vast majority of island universes out there. With each of these 10^{500} states realized in an infinite number of island universes, pure random chance might be enough to "explain" why we happened to evolve where we did.[22]

Three-quarters of this argument is of quite venerable vintage. Back in the seventeenth century, natural philosophers like Bernard Le Bovier de Fontenelle in France and Isaac Newton in England argued that the constants of nature had to be finely tuned in order to support life as we know it. Fontenelle, Newton, and their contemporaries considered such fine-tuning to be scientific proof that God must exist. That is, they took evidence of fine-tuning to be proof of design: of an omnipotent Master Architect at work, designing our universe just for us. Fontenelle and Newton thus were charter members of the "intelligent design" club.[23]

On this last point, Susskind parts company with Fontenelle and Newton. The epigraph to his popular book *The Cosmic Landscape: String Theory and the Illusion of Intelligent Design* (2005) taunts today's adherents of intelligent design. He quotes the response by that famously secular Enlightenment physicist Pierre-Simon de Laplace to Emperor Napoleon, who had asked Laplace where God fit into his view of the cosmos. "Your Highness," Laplace is said to have replied, "I have no need of this hypothesis." Susskind and many of his colleagues have planted their staff firmly in the Darwinians' turf: given enough time and ample possibilities, natural evolution took its cosmic course, and here we are.[24]

Other physicists have reacted to these dramatic claims much as Einstein rebuffed Friedmann's calculations or as Hoyle decried Gamow's creation-talk. Several outspoken critics, including Nobel laureates like Santa Barbara's David Gross, have described the string landscape and anthropic musings to be "dangerous," "disappointing," even an "abdication" of how physicists should approach their craft.[25]

:::

While physicists debate string theory and the landscape, many nonscientists have drawn their own conclusions. One response has been a resurgent biblical literalism. Unlike earlier creationists, however, today's advocates no longer excuse physics and cosmology from their purview. For example, despite a raft of high-precision astrophysical observations that have verified a central tenet of the big bang—that our observable universe is 13.8 billion years old—newly emboldened creationists readily dismiss such timescales. "I wasn't there [at the big bang], and neither were they [cosmologists]," exclaimed accountant and Kansas State Board of Education member John W. Bacon in 1999. Bacon was explaining to journalists why he had voted with a majority of board members to remove the big bang as well as biological evolution from statewide high school curricula. "Based on that," he went on, "whatever explanation they may arrive at is a theory and it should be taught that way." Several other states followed Kansas's lead in the late 1990s.[26] Since then, the teaching bans have ebbed and flowed with various election cycles; the issue is far from finished.

If these education board members need any additional encouragement, they have dozens of new "authoritative" texts to turn to. Books like D. Russell Humphreys's *Starlight and Time: Solving the Puzzle of Distant Starlight in a Young Universe* (1994) have been joined by a raft of recent publications, including Donald DeYoung's *Thousands, Not Billions* (2005), Alex Williams and John Hartnett *Dismantling the Big Bang* (2005), and Jason Lisle's *Taking Astronomy Back: The Heavens Declare Creation* (2006). Several of these authors sport advanced degrees in the physical sciences and are supported by a robust institutional network, including the "Answers in Genesis" ministry, complete with its own lecture circuit and educational museum. Much like the island universes dotting Susskind's string landscape, in other words, today's creationists have carved out a parallel universe all their own. Most of their books have a sales rank on Amazon.com an order of magnitude better than mine.

Alongside biblical literalism, a second response has come from devotees of "intelligent design." While most news coverage of intelligent design has concentrated on fights over biological evolution in the classroom—such as the headline-grabbing legal showdown in Dover, Pennsylvania, during 2005—the issue has cropped up in other surprising places as well. In February 2006, for example, the story broke of a young public-affairs officer at NASA named George Deutsch, who had circulated an internal memorandum stipulating that the word "theory" be appended to "big bang" in all NASA documents, especially educational websites. "The big bang is not proven fact; it is opinion," began Deutsch's memo, a copy of which was leaked to the *New York Times*. "It is not NASA's place, nor should it be, to make a declaration such as this about the existence of the universe that discounts intelligent design by a creator." Although the twenty-four-year-old political appointee was forced out of NASA soon after the memo was leaked, the episode makes plain just how well placed today's advocates of intelligent design have become.[27]

:::

Why have the evolution debates played out so differently in biology and cosmology? Looking back over the past century, two features seem especially salient: pedagogy and prestige.

Unlike biological evolution, the big bang has never been a central part of high school curricula. Modern cosmology draws on material—such as general relativity, let alone string theory—that lies well beyond the scope of secondary school instruction. Thus, whereas Darwinian natural selection has long provided an obvious target for critics of evolution in the classroom, until recently cosmic evolution has been something of a nonissue.

The recent bans on teaching the big bang might not disrupt many lesson plans, but they remain potent symbolically. They signal a sea change in relative prestige. Physicists emerged from the Second World War as national heroes. Their wartime projects had "delivered the goods," and they found themselves fêted like no other group of academics before or since. Biologists enjoyed no such culminating moment at midcentury. Indeed, some historians have argued that American biologists' zeal to use the centennial of Darwin's *Origin of Species* in 1959 to reassert their own cultural importance might have backfired, wakening the sleeping giant of anti-evolution creationists.[28]

Since the end of the Cold War, physicists' cultural standing has changed dramatically. The era of limitless funding came to a sudden halt in the early 1990s. Congress provided a clear indication of the change when it canceled the Superconducting Supercollider in October 1993. For the rest of the decade, federal funding for basic physics continued to slide. The changing fortunes for physics, combined with physicists' own internal divisions and the obviously speculative nature of recent work, have opened the door for concerted criticism and pushback.

Today's critics of cosmology have learned to leverage the power of the internet. I stumbled onto this thriving, wired network a few years ago, after my colleague Alan Guth and I published a review of recent cosmological research in *Science*. About a week after our article appeared, Alan received an email directing him to a rebuttal of our piece, posted on a creationist website. Curious, I checked out the site; one page linked to others, and still more beyond that. I followed dozens of hyperlinks to like-minded "refutations" of the big bang, inflationary cosmology, string theory, and the rest. The sites boasted good graphics and high production values. Single clicks brought me zooming to the homepages of the "Bible-Science Association," the "Creation Science Association," the "Center for Scientific Creation," the "Institute for Creation Research," the "Answers in Genesis" ministry, and dozens of similar groups. I found plenty of sites eager to sell the recent anti-big-bang books, along with DVDs such as

The Privileged Planet, proffering "evidence" of supernatural intelligent design. Separate links revealed detailed "alternative" science lesson plans available for download and offered special nature tours to places like the Grand Canyon, to go sightseeing with specially trained creationist tour guides. (The websites were certainly ecumenical, featuring equal parts biblical literalism and intelligent design.)

Back I clicked to the original website. After quoting extensively from our article, the commentator switched gears. "We had to show you in their own words what these MIT eggheads are saying," it began. (I was impressed: the "egghead" line, at least, suggested keen powers of observation.) "Guth and Kaiser need to take up truck driving. That would get them out of their ivory towers at MIT and into the real world, where they would be forced to look at trees, mountains, weather, ecology, and all the other observable things on our privileged planet that are inexplicable by chance: realities that proclaim *design, purpose, intention*."[29]

Well, I consoled myself: at least someone is still reading *Science*. As for the rest of us in the cosmic evolution business, we'll just have to keep on truckin' . . .

17

No More Lonely Hearts

Back in 1991, science writer Dennis Overbye published a marvelous book, *Lonely Hearts of the Cosmos*. The book traced the development of cosmology — the scientific study of the universe as a whole — during the second half of the twentieth century. The cosmologists in Overbye's book were lonely for two reasons. They included the last remnants of a generation of astronomers who used to sit up all night, alone, under unheated domes, squinting through huge telescopes to catch the faintest glimpses of light from faraway galaxies — a time before large groups of collaborators and automated data collection had become routine. For much of the period that Overbye covered, meanwhile, the field of cosmology hung on the margins of respectability among physicists, a kind of neglected stepchild in the shadow of more flashy fields like high-energy particle physics, with its hulking accelerators and skyrocketing budgets.[1]

Overbye captured the cosmologists' struggles to measure basic features of our universe. Usually their answers could be trusted only to within a factor of two—that is, each measurement carried roughly 100 percent uncertainty. Were galaxies receding from each other at such-and-such a speed, or twice that fast? The answer bore a direct impact on how old our observable universe could be—another key feature that could be pinned down only to within a factor of two. (On the first day of one of my undergraduate courses in the subject, the instructor merrily informed us that we could use the equation "1 = 10," since most quantities of interest had comparable uncertainties. We were not, however, allowed to square that equation.) No wonder cosmologists suffered lonely hearts for so long: those huge uncertainties appeared downright amateurish when compared with triumphs of precision in other branches of physics. For the energy levels of a hydrogen atom, for example, theory and experiment had long since converged, agreeing all the way out to eleven decimal places.

With the basic pace of the universe's expansion so difficult to discern, cosmologists often threw up their hands (or argued at length) over follow-up questions, such as whether the universe's expansion was speeding up or slowing down. The answer to that question would reveal how much stuff the universe contained. A densely packed universe, with lots of matter and energy per cubic meter, should eventually halt its expansion and collapse back onto itself, a big-crunch end to mirror a big-bang beginning. A universe with less matter and energy stuffed into each unit of volume would expand forever, becoming progressively more dilute. Balanced right between, Einstein's equations predicted a Goldilocks solution: some critical amount of stuff per volume for which the rate of expansion would slowly fade but the universe would never recollapse; a gentle, quiet coasting into oblivion. The fate of the entire universe hung on numbers like the present-day expansion rate and the amount of stuff per volume. Yet try as they might—and their efforts were often extraordinarily clever—cosmologists simply could not measure the universe's basic features with requisite confidence or precision.

That began to change, and to change fast, soon after Overbye's *Lonely Hearts* appeared. Indeed, we cosmologists feel a lot less lonely these days. The field is booming, attracting new recruits, fantastic new instruments, and no shortage of exciting new ideas. One often hears talk nowadays of a "golden era" of precision cosmology. In the autumn of 1992—just one year after Overbye's book was published—my fellow undergraduate physics students and I enjoyed a champagne study break with several professors to mark the first release of data from the *Cosmic Background Explorer* (*COBE*) satellite. From its orbit high above the atmosphere, the satellite had measured the first light ever released after the big bang: photons that had been streaming freely throughout the universe since the moment when electrons had begun to combine with protons to form stable, neutral hydrogen atoms, about 380,000 years after our universe began. (Before that moment, ambient temperatures were too high to allow stable hydrogen to form.) From the subtle bumps and wiggles in the distribution of those photons, cosmologists could discern that the temperature of outer space today is just 2.725 degrees above absolute zero and is consistent across the entire sky to about 1 part in 100,000.[2]

The following year, spacewalking astronauts repaired the Hubble Space Telescope, paving the way for further astronomical inquiry unhindered by Earth's atmosphere. Two independent teams used the refurbished Hubble (as well as several large, ground-based telescopes) to study supernovae, cataclysmic explosions from self-destructive stars that can temporarily outshine entire galaxies. Their data, first announced in 1998, suggested the shocking conclusion that our universe's rate of expansion is speeding up. The universe is not just getting bigger; it's getting bigger faster. To reconcile the robust observations with Einstein's relativity, cosmologists were forced to consider some residual energy of empty space—dubbed "dark energy," to signal our ignorance of its origin—whose tendency to stretch space overwhelms the competing gravitational tendency of matter to clump together. Five years after that came the first data from the *Wilkinson Microwave Anisotropy Probe* (*WMAP*), a satellite whose instruments sported resolution thirty times greater than those on *COBE*. Measuring totally different phenomena than the supernova studies, the *WMAP* data confirmed that nearly three-quarters of the energy content of the universe consists of dark energy. In 2013, a different team, using the European Space Agency's *Planck* satellite, reported comparable results, with even greater resolution.

Gone are the days when observable quantities are known only to within a factor of ten or two. Ask those same questions that Overbye's heroes had struggled to answer, and today any cosmologist can rattle off the answers as quickly and confidently as a schoolchild who has mastered the multiplication tables. How quickly are galaxies receding from each other? 67.4 kilometers per second per megaparsec, plus or minus 0.7 percent. How old is our observable universe? 13.80 billion years, plus or minus 0.2 percent. How much matter and energy fills the universe? If one includes the weird and unexpected dark energy in the tally, the total weighs in at precisely the razor-edge critical value, give or take 0.2 percent. When graphing data for some quantities these days, cosmologists must amplify their error bars

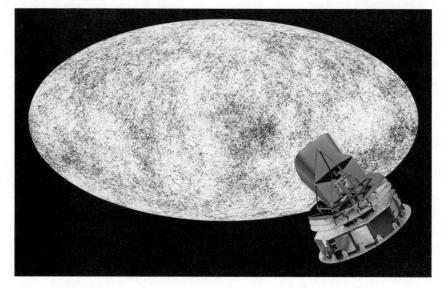

Figure 17.1. Using instruments on the European Space Agency's *Planck* satellite (shown lower right), cosmologists have been able to study subtle patterns in the distribution of energy among photons in the cosmic microwave background radiation, a remnant glow released when the universe was only about 380,000 years old. (*Source*: D. Ducros, © European Space Agency and the *Planck* Collaboration.)

by a factor of four hundred just so the remaining uncertainties can be seen on the page.[3]

Where Overbye had focused on the outsized personalities and human struggles at the heart of cosmology, astrophysicists today often focus on two rather different protagonists: white dwarf stars, especially the type that end their brilliant careers in a particular type of supernova explosion; and the cosmic microwave background radiation, the remnant glow left over from the first formation of stable hydrogen whose pattern has been measured to extraordinary precision by the *WMAP* and *Planck* satellites. Since the mid-2000s, several independent lines of investigation—relying on different instruments that are focused on different physical processes—have more or less converged to give one consistent set of answers. For the first time in human history, scientists can date the age of the cosmos.[4]

Cosmologists thus enjoy an embarrassment of empirical riches these days. We sit immersed in huge data sets bulging with billions of entries, to which experts may devote superlative statistical care. Yet the field has hardly been overrun by stodgy accountants donning green visors. In fact, much of the theoretical activity contributed by professional cosmologists these days looks more bizarre than ever, even absurd; some proposals betray more than a whiff of the circus tent. Sit through a lecture by nearly any cosmologist these days, and before long you are likely to hear phrases like "extra dimensions," "brane-world collisions," "variable equation-of-state quintessence," and "multiverse"—the latter presumed to be an infinitely large container, operating under its own set of physical laws, within which our entire observable universe may be but one tiny bubble.

Of course, neither "bizarre" nor "absurd" imply "incorrect." The long march of cosmology since the Renaissance has been marked by one seemingly preposterous proposal after another, from Copernicus's assertion that the Earth whizzes around the Sun (our own sensations of stillness notwithstanding) to Einstein's suggestion that space and time bend in the presence of matter. Bizarreness, too, is in the eye of the beholder. Even so, cosmologists' collective imagination in recent years has behaved just like an incompressible fluid: try to constrain it within tight quarters (say, by means of precise measurements of observable quantities), and it will squirt out in other directions.

Roger Penrose's recent work is emblematic of the latest imaginative excursions. Now that cosmologists have determined the precise age of our observable universe, Penrose has proposed that all the buzzing, blooming confusion since the big bang has been but a trifle, a finger snap in the longer (perhaps infinite) history of our universe. Rather than presume that the big bang of 13.8 billion years ago was the start of everything, in other words, Penrose has crafted an ambitious model that he calls "conformal cyclic cosmology," or CCC. Penrose suggests that our universe has already passed through innumerable previous instantiations prior to the big bang that started our present epoch, and that it will likely cycle on like this forever more, much as in Friedrich Nietzsche's "eternal recurrence."[5]

The first *C* in Penrose's model — "conformal" — is critical. The most familiar example of a conformal map is the Mercator projection of Earth. Although Earth's surface is roughly spherical, one may represent Earth's features on a flat, two-dimensional map. During the Renaissance, the Flemish cartographer Gerardus Mercator realized that he could stretch and warp the image of Earth's landmasses on his flat map in such a way that he could preserve the angles between shipping routes near crowded ports — information of great interest to navigators. The result was a map that preserved angles and shapes of small objects everywhere on the map but that greatly distorted overall length scales. Hence, Antarctica looms large on a Mercator projection, dwarfing Europe and Asia combined, even though on Earth's actual surface, Europe and Asia together cover nearly four times more area than does Antarctica. More recently, the Dutch artist M. C. Escher featured conformal projections in many of his famous lithographs. (Conformal maps clearly hold special appeal in the Low Countries.)

Physicists and mathematicians have long made use of conformal mappings to simplify a given problem or to view strange solutions from a new vantage point. The technique has proven especially powerful for the study of Einstein's

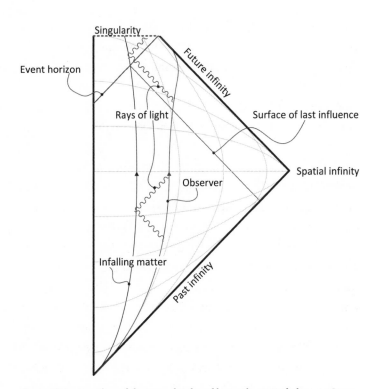

Figure 17.2. A conformal diagram, developed by mathematical physicist Roger Penrose, to study the causal structure of spacetime. Time runs vertically, and light travels along 45-degree diagonals. In this diagram, matter has collapsed into a black hole. A distant observer may send light signals to the infalling matter and receive a response only up until the time when the matter crosses the "event horizon"; from there, as Penrose clarified with diagrams like this, the matter will eventually reach the "singularity" within the black hole, a rupture in spacetime itself. (*Source*: Illustration by Viktor T. Toth.)

general relativity, as a means of gaining leverage on the deformations of spacetime. Penrose made his landmark contributions to mathematical physics back in the mid-1960s by brilliant application of conformal techniques. (In fact, as historian Aaron Wright has documented, Penrose was inspired in part by Escher's playful pictures, which delighted Penrose as a child.)[6] Armed with these powerful graphical methods—now known as "Penrose diagrams"—he demonstrated that a black hole must necessarily lead to a genuine rupture in spacetime, or "singularity." No path, not even a light ray's, can extend beyond some finite limit in the face of a singularity. Penrose's conformal maps proved that the singular behavior was no artifact of this or that coordinate system, nor was it restricted to simple, highly symmetric scenarios.

Penrose has returned to these conformal techniques, now turning them loose on the universe as a whole. He argues that the end of one cosmic epoch, or "aeon," may look quite a lot like the beginning of another—so much so that perhaps they might be stitched together, end on end, into an infinite tower of repeating aeons. During the earliest moments of one aeon, the universe would be hot and dense, just as our observable universe had been right after our big bang. When temperatures are much greater than particles' masses, the particles behave as if they have essentially no mass at all. They zip around at nearly the speed of light, just like photons do. That's critical, because massless particles betray no inherent reference scale—no baseline unit of length or time, no meter stick or calibration clock against which other measures might compare. As far as a photon is concerned, time simply does not flow. A spacetime filled with massless particles would have no inherent scales by which to measure length or time. It would be governed, in other words, by conformal geometry: shapes and angles would have meaning, but overall distances would not.

Remarkably, the end of an aeon might behave in much the same way. As the universe expands and cools after the beginning of one of these cycles, the ambient temperature would drop—looking, for observers within that epoch, just as our own big-bang universe does to us. Massive particles like electrons, protons, hydrogen atoms, and all the rest would gradually lose energy; they would no longer zip around as fast as massless photons do. In that regime, length and time scales would emerge; the symmetries of conformal geometry would be suppressed. The world would behave as ours does today. Pockets of dust would clump and, fueled by the energy of gravitational collapse, ignite into the nuclear reactors we call stars. Eventually, billions of stars would attract each other gravitationally and form tight-knit galaxies. Galaxies themselves would form clusters and superclusters. All the cosmic phenomena that our keenest instruments can observe would unfold: galaxies would recede from each other; certain white dwarf stars would go supernova.

So much for the behavior of the universe after a few tens of billions of years. We know from the supernova measurements and the *WMAP* and *Planck* data that our universe will almost certainly never recollapse upon itself; it should continue to expand forever. So Penrose presses on: what would the universe look like after, say, 10^{100} years, a timescale that makes the present age of our observable universe (not much more than 10^{10} years old) seem positively minuscule? By such late times, nearly all the extant matter would likely have fallen into black holes. Indeed, swarms of black holes would likely have swallowed each other, forming supermassive black holes. But even black holes, it turns out, are far from foolproof containers. Penrose's colleague Stephen Hawking demonstrated in the mid-1970s that black holes should radiate, slowly but surely emitting energy in the form of low-energy light. (This "Hawking radiation" is compatible with Penrose's earlier proofs about singularities. No radiation leaks out from the vicinity of the singularity; the radiation is generated just beyond the boundary of the black hole, known as the "event horizon.") Because of Hawking radiation, black holes behave like cosmic trash compactors: swallowing up massive detritus and ever so slowly seeping that energy back out into the cosmos in the form of massless photons. The process might continue inexorably, until the black holes themselves evaporate. What would be left? A nearly empty universe containing virtually nothing but massless particles—a spacetime, that is, governed once again by conformal geometry.

Penrose, that master of conformal geometry, considers the geometrical similarity of start and end too good to pass up. With more of his mathematical sorcery, he demonstrates how one can smoothly identify the far-future surface of one aeon with the beginning surface of the next, and so on ad infinitum. Sound bizarre? No doubt. Yet Penrose's audacious proposal is in fact rather conservative by today's cosmological standards. For one thing, his model requires just four dimensions of spacetime: one dimension of time and three dimensions of space, the same as in Einstein's physics, let alone Newton's. No need for six or more additional dimensions of space, as superstring theory requires—dimensions that, to hear the string theorists tell, must surely be out there, jutting out at right angles to the height, depth, and breadth that we know and love, yet somehow remaining hidden from view, either because they have mysteriously curled up on themselves and shrunk down to submicroscopic size or because, as luck would have it, we inhabit some strange sausage-like slice (a membrane, or "brane") on which gravity just happens to behave as if there were only three spatial dimensions.[7]

The boundaries between aeons in Penrose's model would betray none of the bizarreries that mark today's ongoing quest for a quantum theory of gravity. Ordinarily, cosmologists expect the superhot, high-energy regimes surrounding a big-bang event to excite quantum fluctuations of spacetime itself. Not only would spacetime behave like a wobbly trampoline, as in Einstein's general relativity, but each tiny unit of space and time would presumably wiggle around in some blur, subject to Heisenberg's uncertainty principle. That might sound exciting, except for the nagging fact that no one has yet produced a workable quantum theory of gravity that might describe the behavior of such quantum-spacetime wiggles. Not to worry, counsels Penrose: spacetime at the aeon boundaries in his model would be perfectly smooth and well behaved, governed by Einstein-like equations. No need to appeal to wild and as-yet-unknown laws of quantum gravity, be it string theory or some other contender.

Amid such delightful flights of cosmological theorizing—bizarre to some, conservative to others—Penrose, too, sits immersed in today's cornucopia of data. Like nearly all cosmologists, he trains his eye on the cosmic microwave background radiation as captured by satellites like *WMAP* and *Planck*. Penrose argues that if his model is correct, then we should be able to see through the boundaries separating various aeons. Subtle features from the previous aeon, before the big bang that started our own, might be imprinted in the cosmic radiation. Those signals would show up as concentric circles in the sky (yet another cyclic feature of his model). For example, a massive black hole might have undergone repeated collisions with comparable objects during the late stages of the previous aeon. Each of those encounters would have generated tremendous bursts of energy, expanding in circles outward from the collision zone. Those ripples would cross the boundary to our own aeon, ultimately appearing as concentric circles of anomalously uniform temperature amid the tiny fluctuations of the cosmic microwave background radiation.

With a collaborator, Penrose released a paper in November 2010 indicating that a close analysis of the *WMAP* data did indeed turn up just such families of concentric circles. Within the space of three days that December, three separate groups reanalyzed the data in the light of Penrose's claim and found no statistical significance. The circles, if really there, were just as likely to show up by chance given the usual understanding of fluctuations in the radiation. Penrose and his colleague quickly offered a response, challenging some of the arcane statistical arguments. The time lag between critique and countercritique shrank from days and weeks to hours. Like the supermassive black hole collisions, Penrose's papers generated an explosive outburst of activity.[8]

Penrose's concentric circles seem not to have withstood the experts' scrutiny and heralded a revolution in cosmology; they are more likely to fade into oblivion like so many UFO enthusiasts' crop circles. Even so, a larger conclusion seems clear. Penrose's zeal to connect his elegant ideas to exacting details of cutting-edge observations captures just what it's like to work in cosmology today. With the field swimming (drowning?) in high-precision data, no longer does it suffice to argue on the basis of mathematical elegance or aesthetic beauty alone. Terabytes of precision data and sophisticated statistical algorithms have become a cure for all those lonely hearts—a Match.com for the cosmos.

18

Learning from Gravitational Waves

A billion years ago (give or take), in a galaxy far, far away, two black holes concluded a cosmic pas de deux. After orbiting each other more and more closely, their mutual gravity tugging each to the other, they finally collided and rapidly merged into one. Their collision released enormous energy—equivalent to about three times the mass of our Sun, if all the Sun's mass were converted to raw energy. The black holes' inspiral, collision, and merger roiled the surrounding spacetime, sending gravitational waves streaming out in every direction at the speed of light.

By the time those waves reached Earth, early in the morning of 14 September 2015, the once-cosmic roar had attenuated to a barely perceptible whimper. Even so, two enormous machines—the kilometers-long detectors of the Laser Interferometer Gravitational-Wave Observatory (LIGO) in Louisiana and Washington State—picked up clear traces of those waves.[1] In October 2017, three longtime leaders of the LIGO effort—Rainer Weiss, Barry Barish, and Kip Thorne—received the Nobel Prize in Physics for this accomplishment.

The discovery was a long time in the making, in human terms as well as astronomical ones. Einstein predicted such waves a century ago, a consequence of general relativity, his elegant theory of gravitation as the warping of spacetime. Yet Einstein's first calculation of gravitational waves was marred by some arithmetic errors (not uncommon, even for Einstein). Before long Einstein and most of the world's experts fell into a decades-long debate over whether such waves should really exist at all. The theorists twisted themselves up in knots: gravitational waves must exist; they might exist; they could not exist; no, indeed, they must exist. Around and around they went. Their spirited arguments—which Daniel Kennefick charts in his fascinating book *Traveling at the Speed of Thought* (2007)—came to sound like the famous vaudeville act: "I can't pay the rent"; "you must pay the rent."[2]

Consensus among theorists emerged slowly over the course of the 1960s. The experts came to agree that gravitational waves really should exist according to the equations of general relativity, and the waves should have specific characteristics. But that still left many questions open. Was general relativity itself a correct description of nature, and could gravitational waves ever be detected?

The question of detection grew at least as murky as the theorists' debates had been. In 1967, physicist Joseph Weber published results of an experiment in which he claimed to have detected such waves—indeed, detected them with a strength about a thousand times greater than what theorists had come to expect. Exciting, puzzling, and quickly controversial, Weber's announcement helped to draw more

Figure 18.1. Beginning in the late 1960s, physicist Joseph Weber claimed to have detected gravitational waves using his large bar detector at the University of Maryland, though he failed to convince fellow experts. (*Source*: Special Collections, University of Maryland Libraries. © 1969 by the University of Maryland.)

attention to the topic at a time when gravitation remained a side issue for most physicists. After greater scrutiny, however, most experts concluded that Weber's early results, and a series of his follow-up experiments, had not actually detected gravitational waves.[3]

Around that time, Rainer Weiss taught an undergraduate course at the Massachusetts Institute of Technology. He assigned as a homework problem the task of investigating a new approach for detecting the waves, different from the one Weber had used. (Students, take note: sometimes homework problems herald Nobel Prize–worthy advances.) What if physicists tried to detect gravitational waves by scrutinizing tiny shifts in the interference patterns of laser beams that had traveled separate paths before recombining at a detector? Gravitational waves should stretch and squeeze a region of space in a particular pattern as they travel through it. Such a disturbance would alter the lengths along which the laser beams traveled, putting the two laser beams out of phase with each other by the time they both reached the detector—a difference that could give rise to a measurable interference pattern. (Two physicists in Moscow, Michaeil Gerstenstein and V. I. Pustovoit, had proposed a similar idea in 1962, though their work was little known in the West at the time.)[4]

The idea was audacious, to say the least. To detect gravitational waves of the expected amplitude using the interference method, physicists would need to be able to distinguish distance shifts of about one part in a thousand-billion-billion. That's like measuring the distance between the Earth and the Sun to within the size of a single atom, while controlling all other sources of vibration and error that could swamp such a minuscule signal. Little wonder that Kip Thorne, another LIGO laureate, assigned a homework problem of his own in his massive 1973 textbook, *Gravitation*—the book he wrote with Charles Misner and John Wheeler—guiding students to the conclusion that interferometry was hopeless as a method for detecting gravitational waves. (Okay, students: maybe some homework problems can be skipped.) After investigating the idea further, however, Thorne became one of the most tenacious advocates for the interferometric approach.[5]

Convincing Thorne was the easy part; attracting funding and students proved much more difficult. Weiss's first proposal to the National Science Foundation, in 1972, was rejected; a follow-up proposal in 1974 received modest funds for a limited feasibility study. He faced considerable difficulty attracting students and convincing his colleagues that the project was worthwhile. As he reported to a program officer at the National Science Foundation in 1976, "Gravitation research, although viewed as fascinating, is considered too hard and unfortunately profitless not only by the average student but also by much of the physics faculty. In short, the atmosphere if not outright hostile to such research is certainly skeptical." Two years later, Weiss observed in another funding proposal that he had "slowly come to the realization that this type of research is best done by secure (possibly foolish) faculty and young post-doctorates of a gambling bent."[6]

As the size of the anticipated project grew — interferometer arms that would stretch kilometers, not meters, decked out with state-of-the-art optics and electronics — so, too, did the projected budget and organization. Sociologist Harry Collins chronicles the next steps in his engrossing study *Gravity's Shadow: The Search for Gravitational Waves* (2004). The project's expanding size and complexity required political mastery as much as physics know-how. Concerns quickly emerged within the scientific community that LIGO would absorb too many resources from other projects, and thus, the proposed laser observatory pitted some astronomers against physicists in quite bitter, public disputes. Meanwhile, the project's leaders learned about interference from more than just laser beams: at one point, efforts to establish one of the two large detectors in Maine foundered on political rivalries and backroom deals among congressional staffers.[7]

Remarkably, the National Science Foundation approved funding for LIGO in 1992; it was (and remains) the largest scientific project ever funded by the foundation. The timing was propitious: the following year, Congress eliminated funding for enormous scientific projects like the Superconducting Supercollider, whose projected price tag was about forty times larger than LIGO's. After the dissolution of the Soviet Union, physicists learned with whiplash speed that Cold War–era justifications for investing in scientific research no longer held predictable sway in Congress. Beginning in the mid-1990s, meanwhile, budgetary brinksmanship entered a whole new era. For more than two decades, planning for long-term projects has had to contend with frequent threats (occasionally realized) of government shutdowns, compounding a budgetary climate focused on short-term projects that can promise quick results. It is difficult to imagine a project like LIGO getting a green light if proposed today.

Yet LIGO demonstrates some benefits of taking a longer view. The project has exemplified a close coupling between research and teaching, well beyond the suggestive homework problems from the early days. Several undergraduates and scores of graduate students were coauthors on the LIGO team's historic article detailing the first direct detection of gravitational waves, published in February 2016. Since 1992, the project has spawned nearly six hundred PhD dissertations in the United States alone, from one hundred universities across thirty-seven states. The studies have ranged well beyond physics, including pathbreaking studies in engineering and software design.[8]

Figure 18.2. Rai Weiss (*center*) congratulated by members of the MIT portion of the LIGO collaboration upon the announcement that Weiss had shared the 2017 Nobel Prize in Physics. Also shown (*from left to right*) are Slawomir Grass, Michael Zucker, Lisa Barsotti, Matthew Evans, David Shoemaker, and Salvatore Vitale. (*Source*: Photograph by Jonathan Wiggs, *Boston Globe*, courtesy of Getty Images.)

LIGO shows what we can accomplish when we fix our eyes on a horizon well beyond a given budget cycle or annual report. By building machines of exquisite sensitivity and training cadres of smart, dedicated young scientists and engineers, we can test our fundamental understanding of nature to unprecedented accuracy. The quest sometimes yields improvements for technologies of everyday life—the GPS navigation system benefited from efforts to test Einstein's general relativity—even though such spin-offs are difficult to forecast.[9] But with patience, tenacity, and luck, we can sometimes catch a glimpse of nature at its most profound.

19

A Farewell to Stephen Hawking

Stephen Hawking delighted in reminding audiences that he was born three hundred years to the day after the death of Galileo, on 8 January 1942. Imagine how Hawking would have reacted could he have known that he would die on 14 March 2018—the hundred and thirty-ninth anniversary of Albert Einstein's birth.

I never got to know Professor Hawking, and yet I found myself mourning his passing as if I had lost a close colleague. Like so many people of my generation, I grew up in a world in which Hawking's name was nearly as familiar as Einstein's. In one way or another I've been grappling with his ideas for my entire career.

Hawking's breakaway best seller, *A Brief History of Time*, appeared in 1988, while I was in high school. By that time I was already immersed in popular books about the wonders of modern physics; the 1980s saw a boom in high-quality, inexpensive paperbacks inviting readers to sample some of the choicest mysteries of quantum theory or admire the austere grandeur of Einstein's general theory of relativity. Yet Hawking's book felt different. It became a sensation, sought by people who had never noticed the raft of earlier books. Hawking's was a book to own and, for some, to read.[1]

A Brief History of Time offered a tour of Hawking's most significant contributions to the field. His earliest work centered on Einstein's general relativity, the work that had thrust Einstein himself into the spotlight decades earlier. According to the theory, space and time are as wobbly as a trampoline. They can bend or distend in the presence of matter and energy. Their curvature, in turn, gives rise to all the phenomena we associate with gravity. Gravitation, according to this line of thinking, is not a force — the outcome of one object tugging on another, as described by Isaac Newton's equations — but a mere consequence of geometry.

Hawking's first major contribution, which he began to develop in his PhD dissertation at the University of Cambridge, was essentially to push Einstein's idea until it broke. What if matter were to become packed so densely within a region of space that spacetime itself ruptured? Hawking, along with his colleague Roger Penrose, clarified the conditions under which solutions to Einstein's equations must devolve into a "singularity," quite literally a point of no return. The Penrose-Hawking singularity theorems (as they came to be known) indicate that under extreme conditions — the centers of black holes, perhaps even the start of our universe itself — spacetime can simply *end*, a cosmic variant of Shel Silverstein's famous sidewalk.

The singularity theorems apply to "classical" spacetimes — that is, to descriptions of space and time that ignore quantum theory, that other great pillar of modern physics. Soon after Hawking completed his PhD in 1966, he began to attack questions at the troublesome boundary between relativity, which describes the largest objects in the cosmos, and quantum theory, which governs matter at the atomic scale. He stumbled upon his most famous finding in the mid-1970s while puzzling through scenarios in which pairs of quantum particles might find themselves near a black hole. If one were to fall in while the other escaped, Hawking suggested, the black hole would appear, to a distant observer, as if it had emitted radiation—precisely what black holes were *not* supposed to allow. In other words, "black holes ain't so black," as he put it in *A Brief History*: they glow. What's more, this radiation could shape a black hole's fate. Over astronomical timescales, the black hole could evaporate, its once-enormous mass seeping out as cosmic static.

These puzzling ideas—equal parts bizarre and exciting—spawned many others, some of which continue to challenge the physics community to this day. Theoretical physicists still grapple with whether information tossed into a black hole could really disappear forever. Must it be scrambled beyond any possible reconstruction, with only a meaningless bath of radiation remaining? Any such process would violate quantum theory, for which a sacrosanct rule is that information can be neither created nor destroyed. Scores of theorists have turned Hawking's arguments around and poked them from every angle, trying to find where the weak joint might lie in the uneasy combination of quantum theory and relativity. Meanwhile, closer to my own research, Hawking's ideas about the big bang and whether our universe could have emerged from an initial singularity continue to animate studies in cosmology.

Famously, Hawking's descriptions of black holes and the big bang came interlaced throughout *A Brief History of Time* with stories of his personal life. He was diagnosed with the

Figure 19.1. Stephen Hawking, shown here in October 1979, produced a string of major insights into the fundamental nature of space, time, and matter while struggling with the degenerative disease ALS. (*Source*: Photograph by Santi Visalli, courtesy of Getty Images.)

degenerative disease amyotrophic lateral sclerosis (ALS) in 1963, at age twenty-one—just as he was beginning his doctoral studies—and was expected to live only a few more years. In his book, Hawking wrote of his determination to carry on, bolstered by meeting Jane Wilde (whom he married in 1965) and soon by the arrival of their three children. Surely these triumphs—the sheer, stubborn fact that Hawking continued to live—drove the fascination with his book just as much as his clever descriptions of warping space-time did.

Propelled by the book's popular success, Hawking rapidly became a full-blown celebrity. He kept up a remarkable travel schedule even as the effects of his ALS became more severe. In October 1999, he visited Harvard for three weeks, just as I was finishing my PhD there. Lines snaked around city blocks once tickets became available for his lectures. (Until then, the only time I had seen lines that long in Cambridge was when *Star Wars: The Phantom Menace* came out, the previous spring.) Between lectures, Hawking and his sizable entourage of nurses and assistants regrouped in the physics building, near my own tiny office. I never dared approach the famous professor myself, but I remember sitting with some of his assistants late into the night, lost amid the buzz and hum. To be in the vicinity of Hawking was to be immersed in an extended web of activity, of people and machines clicking together, a phenomenon documented in the anthropologist Hélène Mialet's fascinating study *Hawking Incorporated* (2012).[2]

Almost two decades later, I had a different sort of encounter with Hawking. During the spring of 2017, several colleagues and I invited him to join a brief essay we were writing, trying to articulate for a broad audience some of the most significant insights that cosmologists had developed and tested about the earliest moments in cosmic history. At first, Hawking objected to the wording of a particular paragraph. My colleagues, who had known him for decades, assumed that he would never change his mind; he could be famously stubborn. Being innocent of that experience, I suggested a modest edit to address his concern. I will never forget the euphoria, the next day, when I received the email from his assistant saying that Hawking liked the edit and would sign on as a coauthor of the essay. Hawking might have generated enduring truths about the cosmos, but at least I could tame a wayward dependent clause or two.[3]

I imagine that Hawking's well-known stubbornness helped keep him alive. He refused to succumb to his disease, outliving his original prognosis by half a century. But the side of him I think of most is his sense of humor, even showmanship. How fitting, I have often thought, that as he lost control over most of the muscles in his face, his expression settled into an impish grin. He seemed media-savvy in a way that Einstein, too, grew to be. As recently as January 2016, for example, Hawking held his own — comedically, if not strategically — with comic actor Paul Rudd in a short film about quantum chess.[4]

I never met Stephen Hawking, but the idea of him — and several of his own ideas — have been with me for much of my life. May his example continue to inspire young people to beat the odds and to ask big, ungainly questions about the universe.

ACKNOWLEDGMENTS

Most of the chapters in this book first appeared as essays, and those earlier versions benefited enormously from talented and patient editors. I am grateful to Sara Abdulla (*Nature*, chapter 5); Thomas Jones (*London Review of Books*, chapter 12); Angela von der Lippe (W. W. Norton, chapter 9); Anthony Lydgate (*New Yorker*, chapters 4 and 19); Paul Myerscough (*London Review of Books*, chapters 1, 6, 10–12, 14, 17); George Musser (*Scientific American*, chapter 13); Corey Powell (*Aeon*, chapter 3); Jamie Ryerson (*New York Times*, chapter 18); Dave Schneider (*American Scientist*, chapter 16); and Michael Segal (*Nautilus*, chapter 2). From this remarkable set I want to single out Paul Myerscough at the *London Review of Books* for special thanks. More than a third of the contents of this book originally appeared as short pieces in the *LRB*. I still remember how nervous I felt, sending in my first piece to Paul a decade ago. (That first essay happens to be the chapter that opens this collection, on Dirac.) Surely a pro like Paul would see through my pretense. On the contrary, he read the piece with care, helped tighten up a few saggy spots—and then kept asking for more. Over the years I have written more than a dozen pieces for the *LRB*, working with Paul on most of them. I learned how to be a physicist and historian in graduate school. To the extent that I have developed any confidence (let alone skill) as a writer, I credit Paul for his quietly persistent coaxing and coaching.

I have enjoyed the great privilege of discussing several of the topics described in these chapters with remarkable colleagues, including Carl Brans, Angela Creager, Joe Formaggio, Peter Galison, Michael Gordin, Alan Guth, Stefan Helmreich, John Krige, Patrick McCray, Erika Milam, Heather Paxson, Lee Smolin, Matt Stanley, Alma Steingart, Kip Thorne, Rai Weiss, Alex Wellerstein, Benjamin Wilson, Anthony Zee, and Anton Zeilinger. Several friends, colleagues, and students shared comments on individual chapters, and it is a pleasure to thank Marc Aidinoff, Marie Burks, Michael Gordin, Yoshiyuki Kikuchi, Robert Kohler, Roberto Lalli, Bernard Lightman, Kathryn Olesko, Alma Steingart, Bruno Strasser, Marga Vicedo, Benjamin Wilson, and Aaron Wright for their helpful feedback.

K. C. Cole, Nell Freudenberger, Alan Lightman, Julia Menzel, and David Singerman read and commented on the full manuscript. The book benefited in countless ways from each of their thoughtful suggestions. I am especially indebted to Alan Lightman for sharing insights with me about his own writing process, including his thoughts on what makes a collection of essays more than the sum of its parts. I have enjoyed Alan's writing for many years—his nonfiction essays and his novels alike—and so I was deeply honored when he kindly agreed to contribute a foreword for this volume. Nell Freudenberger's detailed comments on an early draft of the manuscript really gave the project life and helped to convince me that the collection just might work as a book. Her talents as a novelist are inspiring, and her feedback on my draft felt like a master class in the craft of storytelling. David Singerman's meticulous comments on the penultimate draft were a perfect bookend to Nell's early feedback. He marked up nearly every page. On large matters and small, he helped me keep the most important points in focus.

I am also grateful to Karen Merikangas Darling at the University of Chicago Press for her enthusiasm about the project from the start and for her constructive feedback on the evolving manuscript; and to my literary agent, Max Brockman, for his unstinting encouragement and sound advice.

Lastly I thank my beloved wife, Tracy Gleason, and our children, Ellery and Toby. They have served as sounding boards and helpful critics, never shying away from telling me that a metaphor had gone astray or an analogy for this or that physics concept just wouldn't cut it. A few years ago, when I was away from home at a conference, my phone buzzed. Tracy had texted to ask me what Hawking radiation was; Toby had just asked her about it over dinner. (Typical text message in our family.) I chuckled to myself, wondering how best to get the gist across in some reply texts, surreptitiously tapped out while seated at my conference dinner. Before I began, though, Tracy texted again. "Never mind," she wrote. "Toby just explained it to me." He was ten years old at the time. I don't know what his explanation sounded like; both had forgotten the details by the time I got home from my trip. What mattered was that Toby had been curious, eager to learn more, and not afraid to try to puzzle through a few steps on his own. Ellery has made clear her own preferences for dinnertime discussion topics. Back when she was eight, her birthday present to me was a T-shirt featuring a large picture of her face contorted in her most dramatic, pleading expression, with the caption "Not the Higgs boson again!" But I know she protests too much; she was the one who talked through my then-inchoate analogy about quantum entanglement tests and twins ordering desserts. Hopefully Ellery and Toby will have more questions after reading this book. It is dedicated to them, with all my love.

ABBREVIATIONS

FB   Felix Bloch Papers. Collection number SC303, Stanford
        University Archives, Palo Alto, California.

HAB   Hans A. Bethe Papers. Collection number 14-22-976,
        Division of Rare and Manuscript Collections, Cornell
        University Library, Ithaca, New York.

JAW   John A. Wheeler Papers. American Philosophical
        Society, Philadelphia, Pennsylvania.

KST   Kip S. Thorne Papers. In Professor Thorne's possession,
        California Institute of Technology, Pasadena.

LIS   Leonard I. Schiff Papers. Call number SC220, Stanford
        University Archives, Palo Alto, California.

MIT-AR Massachusetts Institute of Technology Department
        of Physics, Annual Reports. Bound chronologically
        in Reports to the [Institute] President, call number
        T171.M4195, Massachusetts Institute of Technology
        Archives, Cambridge.

PDP   Princeton University Department of Physics records.
        Seeley G. Mudd Manuscript Library, Princeton, New
        Jersey.

RTB   Raymond Thayer Birge Correspondence and Papers.
        Call number 73/79c, Bancroft Library, University of

California–Berkeley. Letters written by Birge are filed chronologically. The items cited here are from boxes 39 and 40; explicit folder titles will not be cited. Letters written to Birge are cited with box and folder titles.

VFW   Victor F. Weisskopf Papers. Collection MC 572, Institute Archives and Special Collections, Massachusetts Institute of Technology, Cambridge.

NOTES

Introduction

1. Handwritten notes between Paul Ehrenfest and Albert Einstein, 25 October 1927, document 10-168, in Einstein Archives, Princeton University.

2. I quoted the exchange in David Kaiser, "Bringing the Human Actors Back on Stage: The Personal Context of the Einstein-Bohr Debate," *British Journal for the History of Science* 27 (1994): 129–52, on 146n89. Also quoted in Jagdish Mehra, *The Solvay Conferences on Physics: Aspects of the Development of Physics since 1911* (Boston: Reidel, 1975), xvii, 152; and Martin Klein, "Einstein and the Development of Quantum Physics," in *Albert Einstein: A Centenary Volume*, ed. Anthony French (Cambridge, MA: Harvard University Press, 1979), 133–51, on 136.

3. See esp. Guido Bacciagaluppi and Antony Valentini, *Quantum Theory at the Crossroads: Reconsidering the 1927 Solvay Conference* (New York: Cambridge University Press, 2009).

4. Thomas Levenson, *Einstein in Berlin* (New York: Bantam, 2003), chap. 23; Paul Ehrenfest to Niels Bohr, May 1931 ("I have completely lost contact"), as quoted in Abraham Pais, *Niels Bohr's Times: In Physics, Philosophy, and Polity* (New York: Oxford University Press, 1991), 409. On Ehrenfest's unsent letter ("enervated and torn," "weary of life") and suicide, see Pais, *Niels Bohr's Times*, 409–11.

5. Dominik Rauch et al., "Cosmic Bell Test Using Random Measurement Settings from High-Redshift Quasars," *Physical Review Letters* 121 (2018): 080403, https://arxiv.org/abs/1808.05966.

6. Samuel Goudsmit to Leonard Schiff, 2 September 1966, in LIS box 4, folder "Physical Review" ("neighborhood grocery store"); and Simon Pasternack to Leonard Schiff, 22 January 1958 and 27 June 1963, in LIS box 4, folder "Physical Review." See also Goudsmit's annual reports in *Physical Review* Annual Reports, Editorial Office of the American Physical Society, Ridge, NY.

7. Samuel Goudsmit, 1956 Annual Report ("too bulky"), 1955 Annual Report ("'six feet' of *The Physical Review*"), and 1963 Annual Report ("psychological limit"), in *Physical Review* Annual Reports. See also W. B. Mann to Samuel Goudsmit, 11 January 1955 ("destruction of the printed word"), in box 79, folder 14, Henry A. Barton Papers, collection number AR20, Niels Bohr Library, American Institute of Physics, College Park, MD; Leonard Loeb to Goudsmit, 19 April 1955, in RTB box 19, folder "Loeb, Leonard Benedict"; and Thomas Lauritsen to Goudsmit, 27 December 1968, in box 12, folder 14, Thomas Lauritsen Papers, California Institute of Technology Archives, Pasadena. See also David Kaiser, "Booms, Busts, and the World of Ideas: Enrollment Pressures and the Challenge of Specialization," *Osiris* 27 (2012): 276–302, on 291–93.

Chapter 1

A version of this essay originally appeared in *London Review of Books* 31 (26 February 2009): 21–22.

1. Abraham Pais, *Niels Bohr's Times: In Physics, Philosophy, and Polity* (New York: Oxford University Press, 1991); David Cassidy, *Uncertainty: The Life and Science of Werner Heisenberg* (San Francisco: W. H. Freeman, 1991); Mary Jo Nye, "Aristocratic Culture and the Pursuit of Science: The de Broglies in Modern France," *Isis* 88 (1997): 397–421; Walter Moore, *Schrödinger: Life and Thought* (New York: Cambridge University Press, 1989); Alexander Dorozynski, *The Man They Wouldn't Let Die* (London: Secker and Warburg, 1966); Charles Enz, *No Time to Be Brief: A Scientific Biography of Wolfgang Pauli* (New York: Oxford University Press, 2002); and Nancy Thorndike Greenspan, *The End of the Certain World: The Life and Science of Max Born* (New York: Basic, 2005).

2. On the development of quantum theory, see esp. Max Jammer, *The Conceptual Development of Quantum Mechanics* (New York: McGraw-Hill, 1966); Olivier Darrigol, *From c-Numbers to q-Numbers: The Classical Analogy in the History of Quantum Theory* (Berkeley: University of California Press, 1992); and Mara Beller, *Quantum Dialogue: The Making of a Revolution* (Chicago: University of Chicago

Press, 1999). On collections of letters, see, e.g., Thomas Kuhn, John Heilbron, Paul Forman, and Lini Allen, *Sources for History of Quantum Physics* (Philadelphia: American Philosophical Society, 1967); K. Przibram, ed., *Letters on Wave Mechanics*, trans. Martin Klein (New York: Philosophical Library, 1967); Albert Einstein, Max Born, and Hedwig Born, *The Born-Einstein Letters* (New York: Macmillan, 1971); Diana K. Buchwald et al., eds., *The Collected Papers of Albert Einstein* (Princeton: Princeton University Press, 1987–); and Wolfgang Pauli, *Wissenschaftlicher Briefwechsel*, ed. Karl von Meyenn, 4 vols. (New York: Springer, 1979–99). On the impact of the Solvay conferences in particular, see esp. Richard Staley, *Einstein's Generation: The Origins of the Relativity Revolution* (Chicago: University of Chicago Press, 2009), chap. 10; and Guido Bacciagaluppi and Antony Valentini, *Quantum Theory at the Crossroads: Reconsidering the 1927 Solvay Conference* (New York: Cambridge University Press, 2009).

3. Graham Farmelo, *The Strangest Man: The Hidden Life of Paul Dirac* (New York: Faber and Faber, 2009). See also Helge Kragh, *Dirac: A Scientific Biography* (New York: Cambridge University Press, 1990). Most biographical details about Dirac in this essay may be found in Farmelo's biography.

4. Ralph Fowler to P. A. M. Dirac, September 1925, as quoted in Farmelo, *Strangest Man*, 83.

5. Werner Heisenberg, "Quantum-Theoretical Re-interpretation of Kinematic and Mechanical Relations," in *Sources of Quantum Mechanics*, ed. B. L. van der Waerden (New York: Dover, 1968), 261–76, on 261. Originally published as "Über quantentheoretische Umdeutung kinematischer und mechanischer Beziehungen," *Zeitschrift für Physik* 33 (1925): 879–93.

6. Quoted in Arthur I. Miller, *Imagery in Scientific Thought* (Boston: Birkhäuser, 1984), 143.

7. Tatsumi Aoyama, Toichiro Kinoshita, and Makiko Nio, "Revised and Improved Value of the QED Tenth-Order Electron Anomalous Magnetic Moment," *Physical Review D* 97 (2018): 036001, https://arxiv .org/abs/1712.06060.

8. Paul Dirac, *The Principles of Quantum Mechanics* (1930), 4th ed. (New York: Oxford University Press, 1982).

9. See, e.g., Kai Bird and Martin Sherwin, *American Prometheus: The Triumph and Tragedy of J. Robert Oppenheimer* (New York: Knopf, 2005); Patricia McMillan, *The Ruin of J. Robert Oppenheimer and the Birth of the Modern Arms Race* (New York: Penguin, 2005); and Richard Polenberg, ed., *In the Matter of J. Robert Oppenheimer: The Secu-*

rity Clearance Hearing (Ithaca, NY: Cornell University Press, 2001). See also Jessica Wang, *American Science in an Age of Anxiety: Scientists, Anticommunism, and the Cold War* (Chapel Hill: University of North Carolina Press, 1999); and David Kaiser, "The Atomic Secret in Red Hands? American Suspicions of Theoretical Physicists during the Early Cold War," *Representations* 90 (Spring 2005): 28–60.

10. In addition to Farmelo, *Strangest Man*, see also Peter Galison, "The Suppressed Drawing: Paul Dirac's Hidden Geometry," *Representations* 72 (Autumn 2000): 145–66.

11. Farmelo, *Strangest Man*, 89.

12. Joshua Wolf Shenk, "Lincoln's Great Depression," *Atlantic*, October 2005; and Frank Manuel, *A Portrait of Isaac Newton* (Cambridge, MA: Harvard University Press, 1968).

13. Farmelo, *Strangest Man*, 425.

14. See, e.g., Ian Hacking, "Making Up People," *London Review of Books* 28 (17 August 2006): 23–26; and Ian Hacking, *Mad Travellers: Reflections on the Reality of Transient Illnesses* (Charlottesville: University Press of Virginia, 1998).

15. See, e.g., Jerome Wakefield, "DSM-5: An Overview of Changes and Controversies," *Clinical Social Work Journal* 41, no. 2 (June 2013): 139–54.

Chapter 2

A version of this essay originally appeared in *Nautilus*, 13 October 2016.

1. See, e.g., the examples analyzed in Robert Crease and Alfred Goldhaber, *The Quantum Moment* (New York: W. W. Norton, 2014), chap. 10.

2. David E. Rowe and Robert Schulmann, eds., *Einstein on Politics* (Princeton: Princeton University Press, 2007); Jimena Canales, *The Physicist and the Philosopher: Einstein, Bergson, and the Debate That Changed Our Understanding of Time* (Princeton: Princeton University Press, 2015), chap. 9; and Walter Moore, *Schrödinger: Life and Thought* (New York: Cambridge University Press, 1989), 249.

3. Albrecht Folsing, *Albert Einstein*, trans. Ewald Osers (New York: Viking Penguin, 1997), chap. 35; and Thomas Levenson, *Einstein in Berlin* (New York: Bantam, 2003), 412–21.

4. Schrödinger to Einstein, 12 August 1933, as quoted in Moore, *Schrödinger*, 275 (see also 267–77).

5. Einstein's letters with Schrödinger are housed in Hebrew Uni-

versity, Jerusalem, and copies are available in the Seeley G. Mudd Manuscript Library at Princeton University. The most cogent analysis of their 1935 exchange remains Arthur Fine, *The Shaky Game: Einstein, Realism, and the Quantum Theory* (Chicago: University of Chicago Press, 1986), chap. 5. I also discuss some of these letters in David Kaiser, "Bringing the Human Actors Back on Stage: The Personal Context of the Einstein-Bohr Debate," *British Journal for the History of Science* 27 (1994): 129–52.

6. Albert Einstein, Boris Podolsky, and Nathan Rosen, "Can Quantum-Mechanical Description of Physical Reality Be Considered Complete?," *Physical Review* 47 (1935): 777–80.

7. Schrödinger to Einstein, 7 June 1935, and Einstein to Schrödinger, 17 June 1935, as quoted and translated in Fine, *Shaky Game*, 66, 68.

8. Einstein to Schrödinger, 19 June 1935, as quoted and translated in Fine, *Shaky Game*, 69.

9. Einstein to Schrödinger, 8 August 1935, as quoted and translated in Fine, *Shaky Game*, 78.

10. Albert Einstein to Paul Ehrenfest, 14 April 1933 ("firmly convinced"), reprinted in Rowe and Schulmann, *Einstein on Politics*, 276; Einstein's remarks on 3 October 1933 in Royal Albert Hall ("lightning flashes"), reprinted in *Einstein on Politics*, 278–81; Einstein to Stephen S. Wise, 6 June 1933 ("secretly re-arming"), reprinted in *Einstein on Politics*, 287–88; on renouncing pacifism, see *Einstein on Politics*, 282–87.

11. Schrödinger to Einstein, 19 August 1935, and Einstein to Schrödinger, 4 September 1935, as quoted and translated in Fine, *Shaky Game*, 82–84.

12. Erwin Schrödinger, "Die gegenwärtige Situation in der Quantenmechanik," *Die Naturwissenschaften* 23 (1935): 807–12, 823–28, 844–49, on 807. An English translation of Schrödinger's essay is available in John Trimmer, "The Present Situation in Quantum Mechanics: A Translation of Schrödinger's 'Cat Paradox' Paper," *Proceedings of the American Philosophical Society* 124 (1980): 323–38.

13. Fine, *Shaky Game*, 80. See also "Dr. Arnold Berliner and *Die Naturwissenschaften*," *Nature* 136 (1935): 506.

14. Schrödinger's diary, July 1933 ("I have already learnt enough"), as quoted in Moore, *Schrödinger*, 272; Max Laue to Fritz London, June 1934, as quoted in Moore, *Schrödinger*, 295; Schrödinger's BBC address from May 1935 ("gallows and stake"), as quoted in Moore, *Schrödinger*, 301–2.

15. Schrödinger to Bohr, 13 October 1935, as quoted in Moore, *Schrödinger*, 313.

16. P. P. Ewald and Max Born, "Dr. Arnold Berliner," *Nature* 150 (1942): 284–85.

17. J. A. Formaggio, D. I. Kaiser, M. M. Murskyj, and T. E. Weiss, "Violation of the Leggett-Garg Inequality in Neutrino Oscillations," *Physical Review Letters* 117 (2016): 050402, http://arxiv.org/abs/1602 .00041.

Chapter 3

A version of this essay originally appeared in *Aeon*, 20 July 2017.

1. See, e.g., Frank Close, *Neutrino* (New York: Oxford University Press, 2010); and Joao Magueijo, *A Brilliant Darkness: The Extraordinary Life and Mysterious Disappearance of Ettore Majorana, the Troubled Genius of the Nuclear Age* (New York: Basic, 2009).

2. Laura Fermi, *Atoms in the Family: My Life with Enrico Fermi* (Chicago: University of Chicago Press, 1954), chaps. 14, 17–19; and Gino Segrè and Betinna Hoerlin, *The Pope of Physics: Enrico Fermi and the Birth of the Atomic Age* (New York: Holt, 2016), chaps. 18–20, 25–27.

3. See esp. Catherine Westfall, Lillian Hoddeson, Paul Henriksen, and Roger Meade, *Critical Assembly: A Technical History of Los Alamos during the Oppenheimer Years, 1943–45* (New York: Cambridge University Press, 1992); and Michael Gordin, *Five Days in August: How World War II Became a Nuclear War* (Princeton: Princeton University Press, 2007).

4. Frederick Reines, "The Neutrino: From Poltergeist to Particle," Nobel Lecture (1995), in *Nobel Lectures, Physics, 1991–1995*, ed. Gösta Ekspong (Singapore: World Scientific, 1997), 202–21.

5. Reines, "Neutrino," 204–5; and "The Reines-Cowan Experiments: Detecting the Poltergeist," *Los Alamos Science* 25 (1997).

6. Reines, "Neutrino"; and Close, *Neutrino*, chap. 3.

7. Frank Close, *Half-Life: The Divided Life of Bruno Pontecorvo, Physicist or Spy* (New York: Basic, 2015), chaps. 1–4. Most biographical details about Pontecorvo in this essay come from Close's *Half-Life*.

8. Simone Turchetti, *The Pontecorvo Affair: A Cold War Defection and Nuclear Physics* (Chicago: University of Chicago Press, 2012); and Close, *Half-Life*. On nuclear patent disputes, see also Alex Wellerstein, "Patenting the Bomb: Nuclear Weapons, Intellectual Property, and Technological Control," *Isis* 99 (2008): 57–87.

9. On Fuchs's wartime espionage, see Robert Williams, *Klaus*

Fuchs: Atom Spy (Cambridge, MA: Harvard University Press, 1987); and David Kaiser, "The Atomic Secret in Red Hands? American Suspicions of Theoretical Physicists during the Early Cold War," *Representations* 90 (Spring 2005): 28–60.

10. Joint Congressional Committee on Atomic Energy, *Soviet Atomic Espionage* (Washington, DC: Government Printing Office, 1951).

11. Close, *Half-Life*, chap. 15.

12. On the journal translations, see David Kaiser, "The Physics of Spin: Sputnik Politics and American Physicists in the 1950s," *Social Research* 73 (Winter 2006): 1225–52.

13. In addition to Close, *Half-Life*, see also Samoil Bilenky, "Bruno Pontecorvo and Neutrino Oscillations," *Advances in High Energy Physics*, 2013, 873236. In his first work on neutrino oscillations, Pontecorvo hypothesized a superposition between a neutrino and an antineutrino; he later modified his model to describe a superposition of two (or more) neutrino flavors.

14. See, e.g., Kaiser, "Atomic Secret in Red Hands?"; and Jessica Wang, *American Science in an Age of Anxiety: Scientists, Anticommunism, and the Cold War* (Chapel Hill: University of North Carolina Press, 1999).

15. Close, *Half-Life*, chap. 17; and Bilenky, "Bruno Pontecorvo."

16. Johanna Miller, "Physics Nobel Prize Honors the Discovery of Neutrino Flavor Oscillations," *Physics Today* 68 (December 2015): 16; and Emily Conover, "Breakthrough Prize in Fundamental Physics Awarded to Neutrino Experiments," *APS News*, 9 November 2015, https://www.aps.org/publications/apsnews/updates/breakthrough.cfm.

17. J. A. Formaggio, D. I. Kaiser, M. M. Murskyj, and T. E. Weiss, "Violation of the Leggett-Garg Inequality in Neutrino Oscillations," *Physical Review Letters* 117 (2016): 050402, http://arxiv.org/abs/1602.00041.

18. Formaggio et al., "Violation of the Leggett-Garg Inequality in Neutrino Oscillations."

Chapter 4

A version of this essay originally appeared in *New Yorker*, 7 February 2017 (online).

1. Albert Einstein to Max Born, 3 March 1947, in *The Born-Einstein Letters, 1916–1955*, ed. Max Born (1971; New York: Macmillan, 2005),

154–55. See also Louisa Gilder, *The Age of Entanglement: When Quantum Physics Was Reborn* (New York: Knopf, 2008).

2. Walter Moore, *Schrödinger: Life and Thought* (New York: Cambridge University Press, 1989).

3. My dessert analogy builds on a similar discussion in Seth Lloyd, *Programming the Universe: A Quantum Computer Scientist Takes on the Cosmos* (New York: Knopf, 2006), 121.

4. Abraham Pais, "Einstein and the Quantum Theory," *Reviews of Modern Physics* 51, no. 4 (December 1979): 863–914, on 907.

5. See esp. Gilder, *Age of Entanglement*; David Kaiser, *How the Hippies Saved Physics: Science, Counterculture, and the Quantum Revival* (New York: W. W. Norton, 2011); Olival Freire, *The Quantum Dissidents: Rebuilding the Foundations of Quantum Mechanics* (New York: Springer, 2014); and Andrew Whitaker, *John Stewart Bell and Twentieth-Century Physics: Vision and Integrity* (New York: Oxford University Press, 2016).

6. See, e.g., Anton Zeilinger, *Dance of the Photons: From Einstein to Quantum Teleportation* (New York: Farrar, Straus, and Giroux, 2010).

7. B. Hensen et al., "Loophole-Free Bell Inequality Violation Using Electron Spins Separated by 1.3 Kilometres," *Nature* 526 (2015): 682–86, https://arxiv.org/abs/1508.05949; M. Giustina et al., "Significant-Loophole-Free Test of Bell's Theorem with Entangled Photons," *Physical Review Letters* 115 (2015): 250401, https://arxiv.org/abs/1511.03190; L. K. Shalm et al., "Strong Loophole-Free Test of Local Realism," *Physical Review Letters* 115 (2015): 250402, https://arxiv.org/abs/1511.03189; W. Rosenfeld et al., "Event-Ready Bell Test Using Entangled Atoms Simultaneously Closing Detection and Locality Loopholes," *Physical Review Letters* 119 (2017): 010402, https://arxiv.org/abs/1611.04604; and M.-H. Li et al., "Test of Local Realism into the Past without Detection and Locality Loopholes," *Physical Review Letters* 121 (2018): 080404, https://arxiv.org/abs/1808.07653.

8. Erwin Schrödinger, "Die gegenwärtige Situation in der Quantenmechanik" (1935), translated in John Trimmer, "The Present Situation in Quantum Mechanics: A Translation of Schrödinger's 'Cat Paradox' Paper," *Proceedings of the American Philosophical Society* 124 (1980): 323–38, on 335.

9. Kaiser, *How the Hippies Saved Physics*.

10. J. Gallicchio, A. Friedman, and D. Kaiser, "Testing Bell's Inequality with Cosmic Photons: Closing the Setting-Independence Loophole," *Physical Review Letters* 112 (2014): 110405, https://arxiv.org/abs/1310.3288.

11. See Zeilinger, *Dance of the Photons*; and Anton Zeilinger, "Light

for the Quantum: Entangled Photons and Their Applications; A Very Personal Perspective," *Physica Scripta* 92 (2017): 072501.

12. T. Scheidl et al., "Violation of Local Realism with Freedom of Choice," *Proceedings of the National Academy of Sciences* 107 (2010): 19708–13, https://arxiv.org/abs/0811.3129.

13. J. Handsteiner et al., "Cosmic Bell Test: Measurement Settings from Milky Way Stars," *Physical Review Letters* 118 (2017): 060401, https://arxiv.org/abs/1611.06985.

14. Handsteiner et al., "Cosmic Bell Test."

15. In each of our experiments, we produced pairs of entangled particles by shining light from a powerful "pump" laser, which had been tuned to emit light of a specific frequency, onto a special piece of material known as a nonlinear crystal. The atomic structure of the crystal is such that when a particle of light (known as a "photon") with a specific frequency enters the crystal, the crystal absorbs the incoming light and emits *pairs* of photons, the sum of whose energy is equal to that carried by the incoming photon. For more details on our experimental setup, see the "supplemental material" available with Handsteiner et al., "Cosmic Bell Test," and with Dominik Rauch et al., "Cosmic Bell Test Using Random Measurement Settings from High-Redshift Quasars," *Physical Review Letters* 121 (2018): 080403, https://arxiv.org/abs/1808.05966.

16. Rauch et al., "Cosmic Bell Test."

Chapter 5

Portions of this essay originally appeared in *Nature* 523 (July 2015): 523–25.

1. See, e.g., Richard Hewlett and Oscar Anderson Jr., *A History of the United States Atomic Energy Commission*, vol. 1, *The New World, 1939–46* (University Park: Pennsylvania State University Press, 1962); Peter Bacon Hales, *Atomic Spaces: Living on the Manhattan Project* (Urbana: University of Illinois Press, 1997); and Henry Guerlac, *Radar in World War II* (1947; New York: American Institute of Physics, 1987).

2. Lincoln Barnett, "J. Robert Oppenheimer," *Life*, 10 October 1949, 120–38, on 121.

3. Joseph Jones, "Can Atomic Energy Be Controlled?," *Harper's*, May 1946, 425–30, on 425 ("dinner party"); and Samuel K. Allison, "The State of Physics, or The Perils of Being Important," *Bulletin of the Atomic Scientists* 6 (January 1950): 2–4, 26–27, on 2–3 ("besieged with requests," "exhibited as lions").

4. On travel to the Shelter Island meeting, see Silvan S. Schweber, *QED and the Men Who Made It: Dyson, Feynman, Schwinger, and Tomonaga* (Princeton: Princeton University Press, 1994), 172–74. On physicists' B-25 flights, see Philip Morse, *In at the Beginnings: A Physicist's Life* (Cambridge, MA: MIT Press, 1977), 247. On correspondence from nonphysicists, see the thick folders of letters in University of California–Berkeley, Department of Physics records, 3:19–21, collection number CU-68, Bancroft Library, University of California–Berkeley; and in Samuel King Allison Papers, 33:3, Special Collections Research Center, University of Chicago Library, Chicago, IL. On the Gallup poll, see Daniel Kevles, *The Physicists: The History of a Scientific Community in Modern America* (1978), 3rd ed. (Cambridge, MA: Harvard University Press, 1995), 391.

5. James B. Conant, "Chemists and the National Defense," *News Edition of the American Chemical Society* 19 (25 November 1941): 1237.

6. On Conant, see esp. James Hershberg, *James B. Conant: Harvard to Hiroshima and the Making of the Nuclear Age* (New York: Knopf, 1993).

7. See David Kaiser, "Shut Up and Calculate!," *Nature* 505 (9 January 2014): 153–55. See also Peter Galison, *Image and Logic: A Material Culture of Microphysics* (Chicago: University of Chicago Press, 1997); and Lillian Hoddeson et al., *Critical Assembly: A Technical History of Los Alamos during the Oppenheimer Years, 1943–1945* (New York: Cambridge University Press, 1993).

8. Henry A. Barton, "A Physicist's War," Bulletin 1 (12 January 1942), in box 11, folder 16, Henry A. Barton Papers, collection number AR20, Niels Bohr Library, American Institute of Physics, College Park, MD.

9. R. J. Havighurst and K. Lark-Horovitz, "The Schools in a Physicist's War," *American Journal of Physics* 11 (April 1943): 103–8, on 103–4 ("New courses in biology"); and Charles K. Morse, "High School Physics and War," *American Journal of Physics* 10 (December 1942): 333–34 ("It is now").

10. Thomas D. Cope et al., "Readjustments of Physics Teaching to the Needs of Wartime," *American Journal of Physics* 10 (October 1942): 266–68.

11. V. R. Cardozier, *Colleges and Universities in World War II* (Westport, CT: Praeger, 1993), 43, 71, 109–11; Donald deB. Beaver and Renee Dumouchel, eds., *A History of Science at Williams* (1995), 2nd ed. (2000), chap. 3, sec. 3, http://www.williams.edu/go/sciencecenter/center/histscipub.html; Karl T. Compton, 1944–45 Annual Report,

in MIT-AR; and John Burchard, *Q.E.D.: M.I.T. in World War II* (New York: Wiley, 1948), chap. 19. See also Deborah Douglas, "MIT and War," in *Becoming MIT: Moments of Decision*, ed. David Kaiser (Cambridge, MA: MIT Press, 2010), 81–102.

12. Henry A. Barton, "A Physicist's War," Bulletin 13 (8 March 1943), in Barton Papers.

13. B. E. Warren, 1942–43 Annual Report, in MIT-AR; unsigned Princeton report from August 1945 in PDP box 1, folder "Report for War Service Bureau"; and Beaver and Dumouchel, *History of Science at Williams*, chap. 3, sec. 3.

14. Richard Rhodes, *The Making of the Atomic Bomb* (New York: Smon and Schuster, 1986).

15. Rebecca Press Schwartz, "The Making of the History of the Atomic Bomb: The Smyth Report and the Historiography of the Manhattan Project" (PhD diss., Princeton University, 2008).

16. Henry DeWolf Smyth, *Atomic Energy for Military Purposes* (Princeton: Princeton University Press, 1946). See also Schwartz, "Making of the History of the Atomic Bomb."

17. Schwartz, "Making of the History of the Atomic Bomb," 67. See also David Kaiser, "The Atomic Secret in Red Hands? American Suspicions of Theoretical Physicists during the Early Cold War," *Representations* 90 (Spring 2005): 28–60.

18. Schwartz, "Making of the History of the Atomic Bomb"; and Michael Gordin, *Five Days in August: How World War II Became a Nuclear War* (Princeton: Princeton University Press, 2007), chap. 7.

19. War Department press release, "State of Washington Site of Community Created by Project" (6 August 1945), in *Manhattan Project: Official History and Documents*, ed. Paul Kesaris, 12 microfilm reels (Washington, DC: University Publications of America, 1977), reel 1, pt. 6.

20. Schwarz, "Making of the History of the Atomic Bomb," chap. 3.

21. US Senate, Special Committee on Atomic Energy, *Essential Information on Atomic Energy* (Washington, DC: Government Printing Office, 1946).

22. Kaiser, "Atomic Secret in Red Hands?" See also Jessica Wang, *American Science in an Age of Anxiety: Scientists, Anticommunism, and the Cold War* (Chapel Hill: University of North Carolina Press, 1999).

23. Hewlett and Anderson, *History of the United States Atomic Energy Commission*, 1:633–34.

24. Emanuel Piore (director of the Physical Sciences Division of the Office of Naval Research), as quoted in Rebecca Lowen, *Cre-*

ating the Cold War University: The Transformation of Stanford (Berkeley: University of California Press, 1997), 106. See also Emanuel Piore, "Investment in Basic Research," *Physics Today* 1 (November 1948): 6–9.

25. Senator B. B. Hickenlooper to David E. Lilienthal, 30 July 1948, reprinted in *Hearings Before the Joint Committee on Atomic Energy, Congress of the United States, Eighty-First Congress, First Session, on Atomic Energy Commission Fellowship Program, May 16, 17, 18, and 23, 1949* (Washington, DC: Government Printing Office, 1949), on 5; Lilienthal's testimony appears on 4. On 1953 Atomic Energy Commission employment statistics, see John Heilbron, "An Historian's Interest in Particle Physics," in *Pions to Quarks: Particle Physics in the 1950s*, ed. Laurie Brown, Max Dresden, and Lillian Hoddeson (New York: Cambridge University Press, 1989), 47–54, on 51.

26. Paul Forman, "Behind Quantum Electronics: National Security as Basis for Physical Research in the United States, 1940–1960," *Historical Studies in the Physical and Biological Sciences* 18 (1987): 149–229.

27. On growth rates for PhD conferrals across fields, see David Kaiser, "Cold War Requisitions, Scientific Manpower, and the Production of American Physicists after World War II," *Historical Studies in the Physical and Biological Sciences* 33 (Fall 2002): 131–59. On broader impacts of the G.I. Bill and shifts within American higher education, see also Stuart W. Leslie, *The Cold War and American Science* (New York: Columbia University Press, 1993); Roger Geiger, *Research and Relevant Knowledge: American Research Universities since World War II* (New York: Oxford University Press, 1993); and Louis Menand, *The Marketplace of Ideas: Reform and Resistance in the American University* (New York: W. W. Norton, 2010).

28. R. C. Gibbs (chair, National Research Council Division of Mathematics and Physical Sciences) to A. G. Shenstone (Princeton physics department chair), 7 August 1950, in PDP box 2, folder "Scientific manpower" ("procedures for utilizing our manpower"); R. C. Gibbs and H. A. Barton, "Proposed Policy Recommendation," three-page memorandum dated 1 August 1950, in the same folder ("very short emergency," "stockpile"); "Supplementary Memorandum for Prospective Graduate Students," mimeographed notice circulated by the University of Rochester physics department (n.d., ca. winter 1951), a copy of which may be found in PDP box 2, folder "Scientific manpower." See also Raymond Birge to R. C. Gibbs, 10 August 1950, in RTB.

29. Henry DeWolf Smyth, "The Stockpiling and Rationing of Scientific Manpower," *Physics Today* 4 (February 1951), 18–24, on 19; and

the Bureau of Labor Statistics report as quoted in Henry Barton, "AIP 1952 Annual Report," *Physics Today* 6 (May 1952): 4–9, on 6. See also George Harrison, "Testimony on Manpower," *Physics Today* 4 (March 1951): 6–7. On the growth rates of training physicists around the world, see also Catherine Ailes and Francis Rushing, *The Science Race: Training and Utilization of Scientists and Engineers, US and USSR* (New York: Crane Russak, 1982); and Burton R. Clark, ed., *The Research Foundations of Graduate Education: Germany, Britain, France, United States, Japan* (Berkeley: University of California Press, 1993).

30. Kaiser, "Cold War Requisitions."

Chapter 6

A version of this essay originally appeared in *London Review of Books* 34 (27 September 2012): 17–18.

1. The report, entitled "On the Transmission of Gamma Rays through Shields," and the accompanying table of integrals, both dated 24 June 1947, are available in HAB box 3, folder 15.

2. On Bethe's early training and career, see Silvan S. Schweber, *In the Shadow of the Bomb: Oppenheimer, Bethe, and the Moral Responsibility of the Scientist* (Princeton: Princeton University Press, 2000); and Silvan S. Schweber, *Nuclear Forces: The Making of the Physicist Hans Bethe* (Cambridge, MA: Harvard University Press, 2012). On Bethe's consulting for the nuclear industry after the war, see esp. Benjamin Wilson, "Hans Bethe, Nuclear Model," in *Strange Stability: Models of Compromise in the Age of Nuclear Weapons* (Cambridge, MA: Harvard University Press, forthcoming). My thanks to Wilson for sharing a draft of his chapter prior to publication.

3. Lorraine Daston, "Enlightenment Calculations," *Critical Inquiry* 21 (1993): 182–202.

4. David Bierens de Haan, *Nouvelles tables d'intégrales définies* (Leiden: P. Engels, 1867).

5. Bush quoted in David Kaiser, *Drawing Theories Apart: The Dispersion of Feynman Diagrams in Postwar Physics* (Chicago: University of Chicago Press, 2005), 84. On the founding of the Institute, see 83–87.

6. Freeman Dyson reported Bethe's advice ("not to expect") in Dyson to his parents, 2 June 1948, quoted in Kaiser, *Drawing Theories Apart*, 86.

7. George Dyson, *Turing's Cathedral: The Origins of the Digital Universe* (New York: Pantheon, 2012). See also Peter Galison, *Image and Logic: A Material Culture of Microphysics* (Chicago: University of Chi-

cago Press, 1997), chap. 8. Most biographical details about John von Neumann presented here come from Dyson's book.

8. David Alan Grier, *When Computers Were Human* (Princeton: Princeton University Press, 2005). See also Richard Feynman, "Los Alamos from Below," in *Surely You're Joking, Mr. Feynman! Adventures of a Curious Character* (New York: W. W. Norton, 1985), 90–118; Jennifer Light, "When Computers Were Women," *Technology and Culture* 40, no. 3 (1999): 455–83; and Matthew Jones, *Reckoning with Matter: Calculating Machines, Innovation, and Thinking about Thinking from Pascal to Babbage* (Chicago: University of Chicago Press, 2016).

9. Dyson, *Turing's Cathedral*, chaps. 10–11.

10. Dyson, *Turing's Cathedral*, chap. 10. See also Paul Ceruzzi, *A History of Modern Computing* (Cambridge, MA: MIT Press, 1998), chap. 1.

11. Dyson, *Turing's Cathedral*, 298.

12. C. P. Snow, *The Two Cultures* (1959; New York: Cambridge University Press, 2001).

13. Morse, as quoted in Dyson, *Turing's Cathedral*, 333.

14. Albert Einstein to Henry Allen Moe, 28 November 1954, as quoted in David Kaiser, "Bringing the Human Actors Back on Stage: The Personal Context of the Einstein-Bohr Debate," *British Journal for the History of Science* 27 (1994): 129–52, on 146.

15. See, e.g., Ceruzzi, *History of Modern Computing*; and Atsushi Akera, *Calculating a Natural World: Scientists, Engineers, and Computers during the Rise of U.S. Cold War Research* (Cambridge, MA: MIT Press, 2007).

Chapter 7

Versions of this essay originally appeared in *Social Research* 73, no. 4 (Winter 2006): 1225–52; and in *Osiris* 27 (2012): 276–302. Reprinted with permission by Johns Hopkins University Press.

1. Benjamin Fine, "Russia Is Overtaking U.S. in Training of Technicians," *New York Times*, 7 November 1954, 1, 80; and "Red Technical Graduates Are Double Those in U.S.," *Washington Post*, 14 November 1955, 21.

2. Robert Shiller, *Irrational Exuberance*, 2nd ed. (Princeton: Princeton University Press, 2005), xvii ("a situation"), 81 ("As prices continue to rise"). See also Donald MacKenzie, *An Engine, Not a Camera: How Financial Models Shape Markets* (Cambridge, MA: MIT Press, 2006), chap. 7.

3. Fine, "Russia Is Overtaking U.S.," 1, 80 ("essential for survival");

Fred M. Hechinger, "U.S. vs. Soviet: Khrushchev's New School Program Points Up the American Lag," *New York Times*, 3 July 1960, E8 ("stockpiles," "cold war of the classrooms"); Nicholas DeWitt, *Soviet Professional Manpower: Its Education, Training, and Supply* (Washington, DC: National Science Foundation, 1955); Alexander Korol, *Soviet Education for Science and Technology* (Cambridge, MA: MIT Press, 1957); and Nicholas DeWitt, *Education and Professional Employment in the USSR* (Washington, DC: National Science Foundation, 1961).

4. DeWitt, *Soviet Professional Manpower*; DeWitt, *Education and Professional Employment*. See also Nicholas DeWitt, "Professional and Scientific Personnel in the U.S.S.R.," *Science* 120 (2 July 1954): 1–4. Biographical details from "Soviet-School Analyst: Nicholas DeWitt," *New York Times*, 15 January 1962, 12. On the founding of Harvard's Russian Research Center, see David Engerman, *Know Your Enemy: The Rise and Fall of America's Soviet Experts* (New York: Oxford University Press, 2009), chap. 2.

5. Korol, *Soviet Education*. Biographical details from Erwin Knoll, "U.S. Schools Must Do More: Red 'Training' Isn't Enough," *Washington Post*, 29 December 1957, E6; and from Donald L. M. Blackmer, *The MIT Center for International Studies: The Founding Years, 1951–1969* (Cambridge, MA: MIT Center for International Studies, 2002), 144, 159 (CIA contract); see also chap. 1 on the center's founding. On the report's reception, see Rowland Evans Jr., "Reds Near 10-1 Engineer Lead," *Washington Post*, 3 November 1957, A14 ("fastidious," "most conclusive study"); Knoll, "U.S. Schools Must Do More" ("solid factual data"); and Harry Schwartz, "Two Ways of Solving a Problem," *New York Times*, 22 December 1957, 132.

6. DeWitt, *Soviet Professional Manpower*, viii, xxvi–xxxviii, 133, 187, 259–61; DeWitt, *Education and Professional Employment*, xxxix, 3, 33, 339, 374, 549–53; and Korol, *Soviet Education*, xi, 391, 400, 407–8 ("unwarranted implications"), 414.

7. On physics curricular comparisons, see Korol, *Soviet Education*, 260–71; DeWitt, *Education and Professional Employment*, 277–80; and Edward M. Corson, "An Analysis of the 5-Year Physics Program at Moscow State University," *Information on Education around the World*, no. 11 (February 1959), published by the Office of Education of the US Department of Health, Education, and Welfare. On the other caveats, see DeWitt, *Soviet Professional Manpower*, 107, 125, 252; Korol, *Soviet Education*, 163, 195, 294, 316, 324, 383–84; and DeWitt, *Education and Professional Employment*, 342, 365, 370, 401.

8. DeWitt, *Soviet Professional Manpower*, 94–95, 158; Korol, *Soviet*

Education, 142–43, 355, 364; and DeWitt, *Education and Professional Employment,* 210, 229–31, 235, 316.

9. DeWitt, *Soviet Professional Manpower,* 168–69; and DeWitt, *Education and Professional Employment,* 341–42.

10. Fine, "Russia Is Overtaking U.S."; "Red Technical Graduates Are Double Those in U.S.," 21. On a 10 November 1955 press conference, see Barbara Barksdale Clowse, *Brainpower for the Cold War: The Sputnik Crisis and the National Defense Education Act of 1958* (Westport, CT: Greenwood Press, 1981), 51. On Allen Dulles and the Joint Congressional Committee on Atomic Energy hearings, see Clowse, *Brainpower,* 25–26. See also Donald Quarles, "Cultivating Our Science Talent: Key to Long-Term Security," *Scientific Monthly* 80 (June 1955): 352–55, on 353; and Lewis Strauss, "A Blueprint for Talent," in *Brainpower Quest,* ed. Andrew A. Freeman (New York: Macmillan, 1957), 223–33, on 226.

11. DuBridge's testimony was quoted in National Science Foundation, 1956 Annual Report, 13, http://www.nsf.gov/pubs. On formation of the national committee, see 1956 Annual Report, 17–19; Howard L. Bevis (chair of the new committee), "America's New Frontier," in Freeman, *Brainpower Quest,* 178–86; and Juan Lucena, *Defending the Nation: U.S. Policymaking to Create Scientists and Engineers from Sputnik to the "War against Terrorism"* (New York: University Press of America, 2005), 40–41. On stocks of scientists and engineers, see DeWitt, *Soviet Professional Manpower,* 255 (see also 223–25).

12. Henry M. Jackson, "Trained Manpower for Freedom," sixteen-page report addressed to the Special NATO Parliamentary Committee on Scientific and Technical Personnel; quotations from 3–4, 6–10. The report is dated 19 August 1957, and its cover marks it for release on 5 September 1957. Jackson's advisory committee included several physicists and mathematicians (such as Richard Courant, Maria Goeppert Mayer, Edward Teller, and John Wheeler), as well as the president of MIT (James Killian), the former president of the National Academy of Sciences (Detlev Bronk), and the president of the Motion Pictures Association (Eric Johnston). A copy of Jackson's report may be found in PDP box 2, folder "Scientific manpower." For more on Jackson's report, see John Krige, "NATO and the Strengthening of Western Science in the Post-Sputnik Era," *Minerva* 38 (2000): 81–108, on 88–93.

13. DeWitt quoted in Homer Bigart, "Soviet Progress in Science Cited," *New York Times,* 1 November 1957, 3; Hoover quoted in Robert Divine, *The Sputnik Challenge: Eisenhower's Response to the Soviet Satellite* (New York: Oxford University Press, 1993), 52–53. On the John-

son hearings, see Clowse, *Brainpower*, 59–60; and Divine, *Sputnik Challenge*, 64–67 (Johnson quotation on 67).

14. "U.S. Sponsored Report Warns on Red Education," *Washington Post*, 28 November 1957, A6; Evans, "Reds Near 10-1 Engineer Lead" ("absolute necessity"); cf. Korol, *Soviet Education*, 398–417 and v–vii (Millikan's preface). The Eisenhower administration's Department of Health, Education, and Welfare released a similar report on 10 November 1957, entitled "Education in Russia," which focused mostly on education at the primary and secondary levels. Eisenhower briefed his cabinet on the report on 8 November 1957, warning them to prepare for a new barrage of questions upon its release. See Clowse, *Brainpower*, 15.

15. On Rabi's 15 October 1957 meeting with Eisenhower, see Clowse, *Brainpower*, 11; Divine, *Sputnik Challenge*, 12–13; and John Rudolph, *Scientists in the Classroom: The Cold War Reconstruction of American Science Education* (New York: Palgrave, 2002), 108. Hutchisson's *Newsweek* quotation in Clowse, *Brainpower*, 19; Hutchisson to AIP Advisory Committee on Education, 4 December 1957, in box 3, folder 3, Elmer Hutchisson Papers, collection number AR30259, Niels Bohr Library, American Institute of Physics, College Park, MD; Teller as quoted in Divine, *Sputnik Challenge*, 15; and Hans Bethe, "Notes for a Talk on Science Education," n.d. (ca. April 1958), on 2, in HAB box 5, folder 4. See also Robert E. Marshak and LaRoy B. Thompson to Congressman Kenneth B. Keating, 22 November 1957, in HAB box 5, folder 4; Samuel K. Allison, "Science and Scientists as National Assets," talk before Chicago Teachers Union, 19 April 1958, on 12–14, in box 24, folder 11, Samuel King Allison Papers, Special Collections Research Center, University of Chicago Library, Chicago, IL; and Frederick Seitz, "Factors concerning Education for Science and Engineering," *Physics Today* 11 (July 1958): 12–15. On media coverage during the National Defense Education Act debates, see Clowse, *Brainpower*, chap. 9; and Divine, *Sputnik Challenge*, 15–16, 92–93, 159–62. Franklin Miller Jr., a physics professor at Kenyon College, warned against "overselling" physics training in the wake of *Sputnik* in a letter to Hutchisson, 2 April 1958, in box 4, folder 23, Hutchisson Papers.

16. Clowse, *Brainpower*, 13, 87 ("Trojan horse"), 91 ("willing to strain"). See also Rudolph, *Scientists in the Classroom*, chaps. 1, 3; and Science Policy Research Division, Legislative Reference Service, Library of Congress, *Centralization of Federal Science Activities*, report to the Subcommittee on Science, Research, and Development of the

House Committee on Science and Astronautics (Washington, DC: Government Printing Office, 1969), 48.

17. On grants and fellowships funded by the act, see Clowse, *Brainpower*, 151–55, 162–67; Divine, *Sputnik Challenge*, 164–66; and Roger Geiger, *Research and Relevant Knowledge: American Research Universities since World War II* (New York: Oxford University Press, 1993), chap. 6. The data on PhDs in the physical sciences and engineering come from National Research Council, *A Century of Doctorates: Data Analysis of Growth and Change* (Washington, DC: National Academy of Sciences, 1978), 12. The issue hardly went away after passage of the National Defense Education Act. See, e.g., Fred M. Hechinger, "Russian Lesson: New Study of Soviet Education Contains Warning to U.S.," *New York Times*, 21 January 1962, 157; and John Walsh, "Manpower: Senate Study Describes How Scientists Fit into Scheme of Things in Red China, Soviet Union," *Science* 141 (19 July 1963): 253–55.

18. Based on data in the series of National Science Foundation reports entitled *American Science Manpower* (Washington, DC: National Science Foundation, 1959–71). On the National Register, see the form letter from Henry A. Barton (director, AIP), dated 16 November 1950, a copy of which may be found in LIS box 1, folder "Amer. Inst. of Physics (AIP)." The so-called "baby boom" played a minor role in driving the rapid burst of training in physical sciences; the demographic bulge of new students began to enter undergraduate studies only in 1964.

19. Office of the Director of Defense Research and Engineering, *Project Hindsight* (Washington, DC: Department of Defense, 1969). See also Daniel Kevles, *The Physicists: The History of a Scientific Community in Modern America* (1978), 3rd ed. (Cambridge, MA: Harvard University Press, 1995), chap. 25; Stuart W. Leslie, *The Cold War and American Science* (New York: Columbia University Press, 1993), chap. 9; Geiger, *Research and Relevant Knowledge*, chaps. 8–9; and Kelly Moore, *Disrupting Science: Social Movements, American Scientists, and the Politics of the Military, 1946–1975* (Princeton: Princeton University Press, 2008), chaps. 5–6. Data for figure 7.2 from National Research Council, *Century of Doctorates*, 12; and National Science Foundation, Division of Science Resources Statistics, *Science and Engineering Degrees, 1966–2001*, report no. NSF 04-311 (Arlington, VA: National Science Foundation, 2004).

20. David Kaiser, "Cold War Requisitions, Scientific Manpower, and the Production of American Physicists after World War II," *His-*

torical Studies in the Physical and Biological Sciences 33 (Fall 2002): 131–59.

21. DeWitt, *Soviet Professional Manpower*, 167–69; and DeWitt, *Education and Professional Employment*, 339–42. A few academics expressed frustration with journalists' roughshod treatment of these and similar studies at the time: George Z. F. Bereday, review of Korol, *Soviet Education*, in *American Slavic and East European Review* 17 (October 1958): 355–59; and Seymour M. Rosen, "Problems in Evaluating Soviet Education," *Comparative Education Review* 8 (October 1964): 153–65. At the time, the Soviet Union had a small number of universities (thirty-three in 1953, forty in 1958) but more than seven hundred technical "institutes," which trained the vast majority of higher-education students. Natural sciences and mathematics were taught *only* at the universities, which, in turn, taught very few students in applied science or engineering. Most of DeWitt's analysis therefore focused on the technical institutes.

22. DeWitt, *Soviet Professional Manpower*, 167–69; and DeWitt, *Education and Professional Employment*, 339–42. US institutions continued to graduate twice as many science students per year as the Soviets through the 1970s: Catherine P. Ailes and Francis W. Rushing, *The Science Race: Training and Utilization of Scientists and Engineers, US and USSR* (New York: Crane Russak, 1982), 65. Of course, DeWitt's and Korol's studies themselves need hardly be taken at face value: Russian expatriates working with CIA funding might not be expected to produce "value-free" studies, especially during such charged times. Nonetheless, any ideological distortions or idiosyncratic choices of emphasis—should these have entered their detailed reports at all—paled in comparison to the ways that various readers treated their efforts. More recent "scientific manpower" projections have proven equally feeble when compared with actual outcomes. See esp. Lucena, *Defending the Nation*, chaps. 4–5; Earl H. Kinmonth, "Japanese Engineers and American Mythmakers," *Pacific Affairs* 64 (Autumn 1991): 328–50; and Michael S. Teitelbaum, *Falling Behind? Boom, Bust, and the Global Race for Scientific Talent* (Princeton: Princeton University Press, 2014).

23. See, e.g., Lucena, *Defending the Nation*, chap. 4; and Kinmonth, "Japanese Engineers and American Mythmakers." Declines in annual PhD conferrals across each category were calculated from data tabulated in the annual National Science Foundation reports, "Science and Engineering Doctorate Awards," 1994–2006, http://www.nsf.gov /statistics/doctorates.

24. David Berliner and Bruce Biddle, *The Manufactured Crisis: Myths, Fraud, and the Attack on America's Public Schools* (New York: Basic, 1995), 95–102; Daniel Greenberg, *Science, Money, and Politics: Political Triumph and Ethical Erosion* (Chicago: University of Chicago Press, 2001), chaps. 8–9; Eric Weinstein, "How and Why Government, Universities, and Industry Create Domestic Labor Shortages of Scientists and High-Tech Workers," unpublished working paper, http://www.nber.orb/~peat/Papers/Folder/Papers/SG/NSF.html; and Lucena, *Defending the Nation*, 104–12, 133. See also Teitelbaum, *Falling Behind?*

25. See esp. Lucena, *Defending the Nation*, chap. 4.

26. Berliner and Biddle, *Manufactured Crisis*; Greenberg, *Science, Money, and Politics*; and Lucena, *Defending the Nation*.

27. Cf. Jeremy Bernstein, *Physicists on Wall Street and Other Essays on Science and Society* (New York: Springer, 2008).

Chapter 8

1. Richard Feynman, Robert Leighton, and Matthew Sands, *The Feynman Lectures on Physics*, 3 vols. (Reading, MA: Addison-Wesley, 1963–65).

2. Feynman, Leighton, and Sands, *Feynman Lectures*, 1:3–5. See also Richard C. M. Jones to Robert B. Leighton, 16 April 1962, and Leighton to Earl Tondreau, 27 March 1963, both in box 1, folder 1, Robert B. Leighton Papers, California Institute of Technology Archives, Pasadena, CA.

3. Leo Bauer to M. W. Cummings, 7 November 1963, in box 1, folder 2, Leighton Papers (emphasis in original).

4. On sales figures, see unsigned memo, ca. November 1968, in box 1, folder 2, Leighton Papers. On the enduring interest in the books, see Robert P. Crease, "Feynman's Failings," *Physics World* 27 (March 2014): 25.

5. Hans Bethe, "30 Years of Physics at Cornell" (ca. 1965), 10, in HAB box 3, folder 21; A. Carl Helmholz, interview with the author, Berkeley, 14 July 1998; and W. C. Kelly, "Survey of Education in Physics in Universities in the United States," 1 December 1962, in box 9, American Institute of Physics, Education and Manpower Division records, collection number AR15, Niels Bohr Library, American Institute of Physics, College Park, MD. See also Victor F. Weisskopf, "Quantum Mechanics," *Science* 109 (22 April 1949): 407–8; and David R. Inglis, "Quantum Theory," *American Journal of Physics*

20 (November 1952): 522–23. See also Stanley Coben, "The Scientific Establishment and the Transmission of Quantum Mechanics to the United States, 1919–32," *American Historical Review* 76 (1971): 442–60; Gerald Holton, "On the Hesitant Rise of Quantum Mechanics Research in the United States," in *Thematic Origins of Scientific Thought*, 2nd ed. (Cambridge, MA: Harvard University Press, 1988), 147–87; and Katherine Sopka, *Quantum Physics in America: The Years through 1935* (New York: American Institute of Physics, 1988).

6. Francis G. Slack, "Introduction to Atomic Physics," *American Journal of Physics* 17 (November 1949): 454.

7. J. Robert Oppenheimer, *Science and the Common Understanding* (New York: Simon and Schuster, 1953), 36–37.

8. See esp. Kai Bird and Martin J. Sherwin, *American Prometheus: The Triumph and Tragedy of J. Robert Oppenheimer* (New York: Knopf, 2005), chaps. 1–2; and Charles Thorpe, *Oppenheimer: The Tragic Intellect* (Chicago: University of Chicago Press, 2006), chap. 2. Oppenheimer quoted in "The Eternal Apprentice," *Time*, 8 November 1948, 70–81, on 70 ("unctuous").

9. Bird and Sherwin, *American Prometheus*, chaps. 2–3.

10. Raymond T. Birge, "History of the Physics Department," 5 vols., in vol. 3, chap. 9, p. 31. Birge's "History" is available in the Bancroft Library, University of California–Berkeley.

11. Bird and Sherwin, *American Prometheus*, 84. See also David Cassidy, "From Theoretical Physics to the Bomb: J. Robert Oppenheimer and the American School of Theoretical Physics," in *Reappraising Oppenheimer: Centennial Studies and Reflections*, ed. Cathryn Carson and David A. Hollinger (Berkeley: University of California Press, 2005), 13–29.

12. Copies of Bernard Peter's notes from Oppenheimer's 1939 Berkeley course (Physics 221) are available in several university libraries, including Caltech and Berkeley. As late as 1947, administrative staff in Berkeley's physics department still fielded repeated requests for copies of Oppenheimer's 1939 lecture notes; see correspondence in box 4, folder 16, University of California–Berkeley, Department of Physics records, collection number CU-68, Bancroft Library, University of California–Berkeley.

13. Felix Bloch's handwritten lecture notes from the mid-1930s are available in FB box 16, folders 13–14. The Caltech communal notebooks were called the "Bone Books" and span 1929–69; they are available in the Caltech archives, Pasadena, CA. See esp. entries by

Sherwood K. Haynes, 6 January 1936, in box 1, vol. 2; and by Martin Summerfield, 10 March 1939, in box 1, vol. 3 (emphasis in original).

14. Edward Condon and Philip Morse, *Quantum Mechanics* (New York: McGraw-Hill, 1929), 1, 2, 7, 10, 17–21, 83; and Edwin Kemble, "The General Principles of Quantum Mechanics, Part 1," *Physical Review Supplement* 1 (1929): 157–215, on 157–58, 175–77. Cf. Arthur Ruark and Harold Urey, *Atoms, Molecules, and Quanta* (New York: McGraw-Hill, 1930); Alfred Landé, *Principles of Quantum Mechanics* (New York: Macmillan, 1937); and Edwin Kemble, *The Fundamental Principles of Quantum Mechanics* (New York: McGraw-Hill, 1937). On reviews, see Paul Epstein, "Quantum Mechanics," *Science* 81 (28 June 1935): 640–41; E. U. Condon, "Quantum Mechanics," *Science* 31 (31 January 1935): 105–6; "Foundations of Physics," *American Physics Teacher* 4 (September 1936): 148; Karl Lark-Horovitz, "Quantum Mechanics," *Science* 87 (1 April 1938): 302; L. H. Thomas, "Quantum Mechanics," *Science* 88 (2 September 1938): 217–19; and "The Fundamental Principles of Quantum Mechanics," *American Physics Teacher* 6 (October 1938): 287–88. My discussion of interwar trends in teaching quantum mechanics within the United States is indebted to several pioneering works, though I find much greater emphasis upon philosophical engagement in the extant teaching materials than has previously been noted. See esp. Silvan S. Schweber, "The Empiricist Temper Regnant: Theoretical Physics in the United States, 1920–1950," *Historical Studies in the Physical Sciences* 17 (1986): 55–98; Nancy Cartwright, "Philosophical Problems of Quantum Theory: The Response of American Physicists," in *The Probabilistic Revolution*, ed. Lorenz Krüger, Gerg Gigerenzer, and Mary S. Morgan (Cambridge, MA: MIT Press, 1987), 2:417–35; Alexi Assmus, "The Molecular Tradition in Early Quantum Theory," *Historical Studies in the Physical and Biological Sciences* 22 (1992): 209–31; and Alexi Assmus, "The Americanization of Molecular Physics," *Historical Studies in the Physical and Biological Sciences* 23 (1992): 1–34.

15. Caltech Bone Book entries: Michael Cohen, 14 May 1953, in box 1, vol. 7; Frederick Zachariasen, 27 May 1953, in box 1, vol. 7; and Kenneth Kellerman, 10 April 1961, in box 1, vol. 9. Copies of the written comprehensive and qualifying exams may be found in LIS box 9, folder "Misc. problems"; in FB box 10, folder 19; in box 3, folder 4, University of California–Berkeley, Department of Physics records, collection number CU-68, Bancroft Library; and in Kelly, "Survey of Education," appendix 19.

16. Raymond T. Birge to E. W. Strong, 30 August 1950 ("disgrace"), in RTB. See also David Kaiser, *How the Hippies Saved Physics: Science, Counterculture, and the Quantum Revival* (New York: W. W. Norton, 2011), 18–19.

17. Jacques Cattell, ed., *American Men of Science*, 10th ed. (Tempe, AZ: Jacques Cattell Press, 1960), s.v. "Nordheim, Dr. L(othar) W(olfgang)." See also William Laurence, "Teller Indicates Reds Gain on Bomb," *New York Times*, 4 July 1954; and John A. Wheeler with Kenneth Ford, *Geons, Black Holes, and Quantum Foam: A Life in Physics* (New York: W. W. Norton, 1998), 202–4.

18. Paul F. Zweifel's handwritten notes on Nordheim's 1950 course at Duke University are available in the Niels Bohr Library, American Institute of Physics; see pp. 8–11, 38–39, 58.

19. Freeman Dyson's handwritten lecture notes from his courses at Cornell (1952) and Princeton (1961), in Professor Dyson's possession, Institute for Advanced Study, Princeton; Enrico Fermi, *Notes on Quantum Mechanics* (1961), 2nd ed. (Chicago: University of Chicago Press, 1995), which reproduces the mimeographed handwritten lecture notes that Fermi distributed to his class at the University of Chicago (1954); Elisha Huggins's handwritten notes on Richard Feynman's course at Caltech (1955), in Professor Huggins's possession, Dartmouth College; Hans Bethe's handwritten lecture notes from Cornell (1957), in HAB box 1, folder 26; Evelyn Fox Keller's handwritten notes on Wendell Furry's course at Harvard (1957), in Professor Keller's possession, MIT; Saul Epstein, "Lecture Notes in Quantum Mechanics" (1958), mimeographed typed lecture notes, available in the University of Nebraska Physics Library, Lincoln; and Edward L. Hill, "Lecture Notes on Quantum Mechanics" (1958), mimeographed typed lecture notes, available in the University of Minnesota Physics Library, Minneapolis. For each course, I was able to estimate enrollments based on PhD conferrals from those departments four and five years later (taking into account average degree-completion times from that era). In those cases for which archival information about actual enrollments remains available, the estimates based on later PhD conferrals matched actual enrollments: Julia Gardner (reference librarian, University of Chicago), email to the author, 16 September 2005; Bethe's course grade sheet available in HAB box 1, folder 26; Roger D. Kirby (chair, Department of Physics, University of Nebraska), email to the author, 15 September 2005; and Mary N. Morley (registrar, Caltech), email to the author, 13 September 2005.

20. Leonard I. Schiff, *Quantum Mechanics* (New York: McGraw-Hill, 1949), xi; and David Bohm, *Quantum Theory* (New York: Prentice Hall, 1951), v.

21. Schiff's handwritten lecture notes from fall 1959, in LIS box 8, folder "Sr. Colloquium 'Relativity and Uncertainty'" (emphasis in original).

22. See reviews of various editions of Schiff's textbook: Weisskopf, "Quantum Mechanics"; Morton Hammermesh, "Quantum Mechanics," *American Journal of Physics* 17 (November 1949): 453–54; Abraham Klein, "Quantum Mechanics," *Physics Today* 23 (May 1970): 70–71; and John Gardner, "Quantum Mechanics," *American Journal of Physics* 41 (1973): 599–600.

23. E. M. Corson, "Quantum Theory," *Physics Today* 5 (February 1952): 23–24 ("rare example").

24. Corson, "Quantum Theory," 23–24 ("concise and well balanced"); and Inglis, "Quantum Theory," 522–23 ("credit of Bohm's book").

25. On Bohm's case, see esp. Ellen Schrecker, *No Ivory Tower: McCarthyism and the Universities* (New York: Oxford University Press, 1986), 135–37, 142–44; F. David Peat, *Infinite Potential: The Life and Times of David Bohm* (Reading, MA: Addison-Wesley, 1997), chaps. 5–8; Russell Olwell, "Physical Isolation and Marginalization in Physics: David Bohm's Cold War Exile," *Isis* 90 (1999): 738–56; Olival Freire, "Science and Exile: David Bohm, the Cold War, and a New Interpretation of Quantum Mechanics," *Historical Studies in the Physical and Biological Sciences* 36 (2005): 1–34; and Shawn Mullet, "Little Man: Four Junior Physicists and the Red Scare Experience" (PhD diss., Harvard University, 2008), chap. 4. On sales of Schiff's book, see Malcolm Johnson to Leonard Schiff, 11 March 1964, in LIS box 9, folder "Schiff: Quantum mechanics." In 1989, Dover Publications issued a reprint of Bohm's 1951 textbook. On Bohm's failed efforts to publish a follow-up textbook, see the correspondence in LIS box 13, folder "Bohm."

26. Edward Gerjuoy, "Quantum Mechanics," *American Journal of Physics* 24 (February 1956): 118.

27. Eyvind Wichmann, "Comments on Quantum Mechanics, by L. I. Schiff (Second Edition)," n.d. (ca. January 1965), in LIS box 9, folder "Schiff: Quantum mechanics" (emphasis in original).

28. Jacques Romain, "Introduction to Quantum Mechanics," *Physics Today* 13 (April 1960): 62 ("avoids philosophical discussion"); D. L. Falkoff, "Principles of Quantum Mechanics," *American Journal of Physics* 20 (October 1952): 460–61 ("philosophically tainted ques-

tions"); and Herman Feshbach, "Clear and Perspicuous," *Science* 136 (11 May 1962): 514 ("musty atavistic to-do").

29. George Uhlenbeck, "Quantum Theory," *Science* 140 (24 May 1963): 886. Statistics on textbook publications come from keyword and call-number searches in the online catalog of the US Library of Congress: http://www.loc.gov. With the aid of several research assistants, I copied every homework problem within this set of textbooks and coded each by whether the problem required students to perform a calculation or to describe a physical effect in short-answer or essay form.

30. Robert Eisberg and Robert Resnick, *Quantum Physics of Atoms, Molecules, Solids, Nuclei, and Particles* (New York: Wiley, 1974), vi, 25, 245, 322; Michael A. Morrison, Thomas L. Estle, and Neal F. Lane, *Quantum States of Atoms, Molecules, and Solids* (Englewood Cliffs, NJ: Prentice-Hall, 1976), xv; Robert Eisberg, email to the author, 7 October 2005; and Robert Resnick, email to the author, 11 October 2005. Enrollment changes calculated from data in the annual American Institute of Physics graduate-student surveys, 1961–75, available in American Institute of Physics, Education and Manpower Division records, collection number AR15, Niels Bohr Library. One may find a similar shift in the types of homework problems included in textbooks on quantum mechanics aimed at undergraduates. Compare, e.g., A. P. French, *Principles of Modern Physics* (New York: Wiley, 1958), with A. P. French and Edwin F. Taylor, *An Introduction to Quantum Physics* (New York: W. W. Norton, 1978). Moreover, one finds that textbooks written by physicists in other countries fit the same pattern regarding correlations between pedagogical style and enrollments. Textbooks on quantum mechanics after the Second World War by authors in the United Kingdom and the Soviet Union, which each experienced surges in physics enrollments, included only a small proportion of discussion-style homework problems until the enrollments fell. Physicists in other European countries, such as France, West Germany, and Austria — which did not experience a large spike in physics enrollments after the war — continued to publish textbooks similar to the interwar models, with lengthy chapters on philosophical interpretations of the quantum-mechanical formalism.

31. Raymond T. Birge to E. B. Roessler, 29 November 1952 ("subjects that are not trivial"); Birge to K. T. Bainbridge, 11 February 1953 ("not the sort of work"); and Birge to Alfred Kelleher, 3 November 1954, all in RTB. On the other promotion case, see Birge to Dean A. R. Davis, 9 April 1951, in RTB.

32. On incoming graduate-student enrollments in Stanford's physics department, see faculty meeting minutes, 12 January 1970, in FB box 12, folder 1; and unsigned memo, "Graduate Enrollment and Projection," 2 February 1972, in FB box 12, folder 8. On fears of becoming a "factory," see Paul Kirkpatrick, memo to Stanford physics department faculty, 19 January 1956, in FB box 10, folder 2; and Ed Jaynes, memo to department faculty, 27 April 1956, in FB box 10, folder 3. See also the anonymous memos on comprehensive exam results, 7 and 14 April 1956, in FB box 10, folder 3 ("Rather limited knowledge"); 2 February 1958, in FB box 10, folder 8; W. E. Meyerhof, minutes of Graduate Study Committee meeting, 4 November 1959, in FB box 10, folder 12; and Felix Bloch, "Oral Examinations" memorandum, 9 May 1961, in FB box 10, folder 16.

33. "Faculty Skit 1963," available in University of Illinois at Urbana–Champaign, Department of Physics, Faculty Skits, 1963–73, deposited in the Niels Bohr Library.

34. A. L. Fetter memo to department faculty, 28 February 1972, in FB box 12, folder 8; and comprehensive exam (21–22 September 1972) in FB box 12, folder 10. On the new seminar, see W. E. Meyerhof, memo to Stanford's physics graduate students, 29 September 1972, in FB box 12, folder 10.

35. On Feynman's "Physics X" course, see James Gleick, *Genius: The Life and Science of Richard Feynman* (New York: Pantheon, 1992), 398–99.

36. Kaiser, *How the Hippies Saved Physics*, 19–20.

Chapter 9

Versions of this essay appeared in David Kaiser, *How the Hippies Saved Physics: Science, Counterculture, and the Quantum Revival* (New York: W. W. Norton, 2011), chap. 7; and in *Isis* 103 (2012): 126–38.

1. See also David Kaiser and W. Patrick McCray, eds., *Groovy Science: Knowledge, Innovation, and American Counterculture* (Chicago: University of Chicago Press, 2016).

2. Fritjof Capra, *The Tao of Physics: An Exploration of the Parallels between Modern Physics and Eastern Mysticism* (Boulder, CO: Shambhala, 1975).

3. Fritjof Capra, *Uncommon Wisdom: Conversations with Remarkable People* (New York: Simon and Schuster, 1988), 22–25. The Santa Cruz physicist who invited Capra was Michael Nauenberg; see Nauenberg

interview with Randall Jarrell, 12 July 1994, on 37. Transcript available at http://physics.ucsc.edu/~michael/oral2.pdf.

4. Capra, *Uncommon Wisdom*, 23 ("schizophrenic life"), 27 (on Alan Watts). On Watts's connections with Esalen, see Jeffrey Kripal, *Esalen: America and the Religion of No Religion* (Chicago: University of Chicago Press, 2007), 59, 73, 76, 99, 121–25.

5. Capra, *Uncommon Wisdom*, 34. Capra opened *The Tao of Physics* (11) by recounting his "Dance of Shiva" experience on the beach.

6. Capra, *Uncommon Wisdom*, 34.

7. Fritjof Capra to Victor F. Weisskopf, 12 November 1972, in VFW box NC1, folder 26.

8. Ibid. On Weisskopf's career, see David Kaiser, "Weisskopf, Victor Frederick," in *New Dictionary of Scientific Biography* (New York: Scribner's, 2007), 7:262–69; and Victor F. Weisskopf, *The Joy of Insight: Passions of a Physicist* (New York: Basic, 1991). The oft-stolen textbook was J. M. Blatt and V. F. Weisskopf, *Theoretical Nuclear Physics* (New York: John Wiley, 1952).

9. Capra to Weisskopf, 11 January 1973 (quotations), and Capra to Weisskopf, 23 March 1973, both in VFW box NC1, folder 26.

10. Weisskopf to Capra, 19 April 1973, in VFW box NC1, folder 26.

11. Capra, *Uncommon Wisdom*, 44–45 ("rather hard-headed"), 53–54. Capra's early essays include Fritjof Capra, "The Dance of Shiva: The Hindu View of Matter in the Light of Modern Physics," *Main Currents in Modern Thought* 29 (September–October 1972): 15–20; and Fritjof Capra, "Bootstrap and Buddhism," *American Journal of Physics* 42 (January 1974): 15–19. On Chew's bootstrap program, see David Kaiser, *Drawing Theories Apart: The Dispersion of Feynman Diagrams in Postwar Physics* (Chicago: University of Chicago Press, 2005), chaps. 8–9.

12. Capra, *Uncommon Wisdom*, 46; and Judith Appelbaum, "Paperback Talk: A Science with Mass Appeal," *New York Times*, 20 March 1983, 39–40. On Shambhala Press, see also Sam Binkley, *Getting Loose: Lifestyle Consumption in the 1970s* (Durham, NC: Duke University Press, 2007), 120–22.

13. Capra to Weisskopf, 7 May 1976, and Weisskopf to Capra, 21 June 1976 (quotations), both in VFW box NC1, folder 26.

14. On sales, see Capra to Weisskopf, 8 July 1976, in VFW box NC1, folder 26; and Appelbaum, "Paperback Talk." On subsequent editions and translations, see the full list at http://www.fritjofcapra .net (accessed 12 June 2008).

15. Several years later, two comparative-religion scholars scoffed that Capra's book "seemed to misinterpret Asian religions and cultures on almost every page": Andrea Grace Diem and James R. Lewis, "Imagining India: The Influence of Hinduism on the New Age Movement," in *Perspectives on the New Age*, ed. James R. Lewis and J. Gordon Melton (Albany: State University of New York Press, 1992), 48–58, on 49.

16. Karen de Witt, "Quantum Theory Goes East: Western Physics Meets Yin and Yang," *Washington Post*, 9 July 1977, C1 ("Tall and slim"); and Capra, *Tao of Physics*, quotations on 307.

17. Capra, *Tao of Physics*, 19, 25, 141.

18. Capra, *Tao of Physics*, 160 ("this notion," coat of arms); see also 114–15 and chaps. 11–13.

19. Jack Miles, "A Whole-Earth Scientific Order for the Future," *Los Angeles Times*, 4 April 1982, N8 ("amazingly well"); Jonathan Westphal, in Christopher Clarke, Frederick Parker-Rhodes, and Jonathan Westphal, "Review Discussion: *The Tao of Physics* by F. Capra," *Theoria to Theory* 11 (1978): 287–300, on 294 ("Capra is clearly in earnest"); and Abner Shimony, "Meeting of Physics and Metaphysics," *Nature* 291 (4 June 1981): 435–36, on 436. For other reviews, see George B. Kauffman, "The Tao of Physics," *Isis* 68 (1977): 460–61; A. Dull, "The Tao of Physics," *Philosophy East and West* 28 (1978): 387–90; D. White, "The Tao of Physics," *Contemporary Sociology* 8 (1979): 586–87; Sal P. Restivo, "Parallels and Paradoxes in Modern Physics and Eastern Mysticism, Part I: A Critical Reconnaissance," *Social Studies of Science* 8 (1978): 143–81; Sal P. Restivo, "Parallels and Paradoxes in Modern Physics and Eastern Mysticism, Part II: A Sociological Perspective on Parallelism," *Social Studies of Science* 12 (1982): 37–71; and Robert K. Clifton and Marilyn G. Regehr, "Toward a Sound Perspective on Modern Physics: Capra's Popularization of Mysticism and Theological Approaches Reexamined," *Zygon* 25 (March 1990): 73–104.

20. Isaac Asimov, "Scientists and Sages," *New York Times*, 27 July 1978, 19; and Jeremy Bernstein, "A Cosmic Flow," *American Scholar* 48 (Winter 1978–79): 6–9.

21. Capra, *Tao of Physics*, 25; and V. N. Mansfield, "The Tao of Physics," *Physics Today* 29 (August 1976): 56.

22. Capra to Weisskopf, 8 July 1976, in VFW box NC1, folder 26; David Harrison, "Teaching *The Tao of Physics*," *American Journal of Physics* 47 (September 1979): 779–83, on 779 ("This leads naturally"); and Eric Scerri, "Eastern Mysticism and the Alleged Parallels with

Physics," *American Journal of Physics* 57 (August 1989): 687–92, on 688 ("Anyone involved"). Jack Sarfatti likewise adopted Capra's book as a textbook for one of his popular seminars on science and religion, run by the Physics/Consciousness Research Group: Jack Sarfatti, "Physics/Consciousness Program, De Anza-Foothill College, Spring Quarter 1976," on 4–5, in JAW, Sarfatti folders.

23. Clifton and Regehr, "Toward a Sound Perspective," 73–74.

24. Pedagogical critiques include Donald H. Esbenshade Jr., "Relating Mystical Concepts to Those of Physics: Some Concerns," *American Journal of Physics* 50 (March 1982): 224–28; and Scerri, "Eastern Mysticism." Cf. David Harrison, "Comment on 'Relating Mystical Concepts to Those of Physics'" (letter to the editor), *American Journal of Physics* 50 (October 1982): 873 ("most of these students"); and David Harrison, email to the author, 3 July 2007 ("bums in the seats").

25. Harrison, "Comment on 'Relating Mystical Concepts to Those of Physics,'" 873–74; David Harrison, "Bell's Inequality and Quantum Correlations," *American Journal of Physics* 50 (September 1982): 811–16; Nick Herbert, email to the author, 16 April 2008; and Harrison, email to the author, 17 April 2008. The first quantum mechanics textbook to include any material on Bell's theorem was J. J. Sakurai, *Modern Quantum Mechanics* (Menlo Park, CA: Benjamin Cummings, 1985), 223–32; see L. E. Ballentine, "Resource Letter IQM-2: Foundations of Quantum Mechanics since the Bell Inequalities," *American Journal of Physics* 55 (September 1987): 785–92, on 787.

26. Several reviewers highlighted this "ideological" use of Capra's book: physicists could use it as a hedge against antiscientific sentiments of the day. See Kauffman, "Tao of Physics," 461; Restivo, "Parallels and Paradoxes, Part II," 39, 43, 45, 47, 53; and Scerri, "Eastern Mysticism," 688.

27. Kaiser, *How the Hippies Saved Physics*, 164–69, 276–83, 312–13.

28. See also Cyrus C. M. Mody, "Santa Barbara Physicists in the Vietnam Era," in Kaiser and McCray, *Groovy Science*, 70–106.

Chapter 10

Portions of this essay originally appeared in *London Review of Books* 31 (17 December 2009): 19–20; and in *London Review of Books*, 22 March 2010 (online).

1. My internship was with a portion of the Solenoidal Detector Collaboration.

2. David Kaiser, "Distinguishing a Charged Higgs Signal from a Heavy W_R Signal," *Physics Letters B* 306 (1993): 125–28.

3. Daniel Kevles, "Preface, 1995: The Death of the Superconducting Super Collider in the Life of American Physics," in *The Physicists: The History of a Scientific Community in Modern America* (1978), 3rd ed. (Cambridge, MA: Harvard University Press, 1995), ix–xlii; Michael Riordan, Lillian Hoddeson, and Adrienne Kolb, *Tunnel Visions: The Rise and Fall of the Superconducting Super Collider* (Chicago: University of Chicago Press, 2015); and Joseph Martin, *Solid State Insurrection: How the Science of Substance Made American Physics Matter* (Pittsburgh, PA: University of Pittsburgh Press, 2018), chap. 9.

4. John Heilbron and Robert Seidel, *Lawrence and His Laboratory* (Berkeley: University of California Press, 1989), 135, 235–40, 478–84.

5. Recounted in Robert Serber with Robert P. Crease, *Peace and War: Reminiscences of a Life on the Frontiers of Science* (New York: Columbia University Press, 1998), 148.

6. Peter Westwick, *The National Labs: Science in an American System, 1947–1974* (Cambridge, MA: Harvard University Press, 2003).

7. Richard Hewlett and Francis Duncan, *A History of the United States Atomic Energy Commission*, vol. 2, *Atomic Shield, 1947–1952* (University Park: Pennsylvania State University Press, 1969), 249–50; Robert Seidel, "Accelerating Science: The Postwar Transformation of the Lawrence Radiation Laboratory," *Historical Studies in the Physical Sciences* 13 (1983): 375–400, on 394–97; and Henry DeWolf Smyth as quoted in Robert Seidel, "A Home for Big Science: The Atomic Energy Commission's Laboratory System," *Historical Studies in the Physical Sciences* 16 (1986): 135–75, on 148 ("big groups of scientists").

8. Joseph Platt to Paul McDaniel memorandum, 27 July 1961, as quoted in Robert Seidel, "The Postwar Political Economy of High-Energy Physics," in *Pions to Quarks: Particle Physics in the 1950s*, ed. Laurie Brown, Max Dresden, and Lillian Hoddeson (New York: Cambridge University Press, 1989), 497–507, on 502.

9. Fermilab founding director Robert Wilson's 1969 congressional testimony is quoted in Lillian Hoddeson, Adrienne Kolb, and Catherine Westfall, *Fermilab: Physics, the Frontier, and Megascience* (Chicago: University of Chicago Press, 2008), 13–14.

10. See, e.g., Steven Weinberg, *Dreams of a Final Theory* (New York: Pantheon, 1993); Leon Lederman with Dick Teresi, *The God Particle* (Boston: Houghton Mifflin, 1993); cf. Martin, *Solid State Insurrection*,

chap. 9. On the early years of the SSC project, see Riordan, Hoddeson, and Kolb, *Tunnel Visions*, chaps. 2–3.

11. Geoff Brumfiel, "LHC Sees Particles Circulate Once More," *Nature*, 23 November 2009, doi:10.1038/news.2009.1104.

12. Ian Sample, "Totally Stuffed: CERN's Electrocuted Weasel to Go on Display," *Guardian*, 27 January 2017.

13. See, e.g., Dominique Pestre and John Krige, "Some Thoughts on the Early History of CERN," in *Big Science: The Growth of Large-Scale Research*, ed. Peter Galison and Bruce Hevly (Stanford: Stanford University Press, 1992), 78–99.

Chapter 11

Portions of this essay originally appeared in *London Review of Books* 31 (17 December 2009): 19–20.

1. Murray Gell-Mann, "A Schematic Model of Baryons and Mesons," *Physics Letters* 8 (1964): 214–15. Preprints of Zweig's 1964 papers are available on the CERN website: George Zweig, "An SU$_3$ Model for Strong Interaction Symmetry and Its Breaking," version 1 (dated 17 January 1964), http://cds.cern.ch/record/352337/files; and George Zweig, "An SU$_3$ Model for Strong Interaction Symmetry and Its Breaking," version 2 (dated 21 February 1964), http://cds.cern.ch /record/570209/files. See also Michael Riordan, *The Hunting of the Quark: A True Story of Modern Physics* (New York: Simon and Schuster, 1987).

2. See, e.g., Lillian Hoddeson, Laurie Brown, Michael Riordan, and Max Dresden, eds., *The Rise of the Standard Model* (New York: Cambridge University Press, 1997).

3. MIT physicist Frank Wilczek has described this process as a migration from "c-world to p-world," an almost alchemical transformation of concepts into physical stuff in the world around us. See Frank Wilczek, *The Lightness of Being: Mass, Ether, and the Unification of Forces* (New York: Basic, 2008), 186.

4. Peter Galison, *How Experiments End* (Chicago: University of Chicago Press, 1987), chap. 4.

5. For an accessible account, see Wilczek, *Lightness of Being*.

6. Adrian Cho, "At Long Last, Physicists Calculate the Proton's Mass," *Science*, 21 November 2008.

Chapter 12

Portions of this essay originally appeared in *London Review of Books* 33 (25 August 2011): 20; in *London Review of Books*, 6 July 2012 (online); and in *Huffington Post*, 10 February 2014.

1. Feynman quoted in Michael Riordan, *The Hunting of the Quark: A True Story of Modern Physics* (New York: Simon and Schuster, 1987), 152.

2. Leon Lederman with Dick Teresi, *The God Particle* (Boston: Houghton Mifflin, 1993).

3. See, e.g., Lillian Hoddeson, Laurie Brown, Michael Riordan, and Max Dresden, eds., *The Rise of the Standard Model* (New York: Cambridge University Press, 1997), chap. 28; and Sean Carroll, *The Particle at the End of the Universe: How the Hunt for the Higgs Boson Leads Us to the Edge of a New World* (New York: Dutton, 2012), chap. 8.

4. F. Englert and R. Brout, "Broken Symmetry and the Mass of Gauge Vector Mesons," *Physical Review Letters* 13 (1964): 321–23; Peter Higgs, "Broken Symmetries and the Masses of Gauge Bosons," *Physical Review Letters* 13 (1964): 508–9; and G. S. Guralnik, C. R. Hagen, and T. W. B. Kibble, "Global Conservation Laws and Massless Particles," *Physical Review Letters* 13 (1964): 585–87.

5. Frank Wilczek, "Thanks, Mom! Finding the Quantum of Ubiquitous Resistance," *NOVA: The Nature of Reality* (blog), 4 July 2012, http://www.pbs.org/wgbh/nova/blogs/physics/2012/07/thanks -mom; and John Ellis, "What Is the Higgs Boson?," https://videos .cern.ch/record/1458922.

6. See, e.g., Carroll, *Particle at the End of the Universe*.

7. See Peter Galison, *Image and Logic: A Material Culture of Microphysics* (Chicago: University of Chicago Press, 1997).

8. John Gunion, Howard Haber, Gordon Kane, and Sally Dawson, *The Higgs Hunter's Guide* (New York: Addison-Wesley, 1990).

9. A video of the 13 December 2011 CERN press conference is available at https://videos.cern.ch/record/1406043.

10. G. Aad et al. (ATLAS Collaboration), "Observation of a New Particle in the Search for the Standard Model Higgs Boson with the ATLAS Detector at the LHC," *Physics Letters B* 716 (2012): 1–29; and S. Chatrchyan et al. (CMS Collaboration), "Observation of a New Boson at a Mass of 125 GeV with the CMS Experiment at the LHC," *Physics Letters B* 716 (2012): 30–61.

11. Based on searches in titles, abstracts, and keywords for "Higgs"

and/or "electroweak symmetry breaking" in the Thomson Reuters Web of Knowledge database (formerly the Science Citation Index).

12. Matthew Strassler, *Of Particular Significance* (blog), 4 July 2012, https://profmattstrassler.com/2012/07/04/the-day-of-the -higgs.

Chapter 13

Versions of this essay originally appeared in *Social Studies of Science* 36 (August 2006): 533–64; and in *Scientific American* 296 (June 2007): 62–69. Reprinted with permission. Copyright 2007, *Scientific American*, a Division of Springer Nature America, Inc. All rights reserved.

1. F. L. Bezrukov and M. E. Shaposhnikov, "The Standard Model Higgs Boson as the Inflaton," *Physics Letters B* 659 (2008): 703, https://arxiv.org/abs/0710.3755.

2. E.g., D. I. Kaiser, "Constraints in the Context of Induced Gravity Inflation," *Physical Review D* 49 (1994): 6347–53, https://arxiv.org/abs /astro-ph/9308043; D. I. Kaiser, "Induced-Gravity Inflation and the Density Perturbation Spectrum," *Physics Letters B* 340 (1994): 23–28, https://arxiv.org/abs/astro-ph/9405029; and D. I. Kaiser, "Primordial Spectral Indices from Generalized Einstein Theories," *Physical Review D* 52 (1995): 4295–4306, https://arxiv.org/abs/astro-ph/9408044.

3. Rates of preprints derived from data available at https://arxiv .org (accessed 24 October 2018).

4. See, e.g., Max Jammer, *Concepts of Mass in Classical and Modern Physics* (Cambridge, MA: Harvard University Press, 1961); and Max Jammer, *Concepts of Mass in Contemporary Physics and Philosophy* (Princeton: Princeton University Press, 2000).

5. For an accessible introduction to Mach's principle, see Clifford Will, *Was Einstein Right? Putting General Relativity to the Test*, 2nd ed. (New York: Basic, 1993), 149–53. See also Julian Barbour and Herbert Pfister, eds., *Mach's Principle: From Newton's Bucket to Quantum Gravity* (Boston: Birkhäuser, 1995). On Mach's influences on Einstein, see esp. Gerald Holton, "Mach, Einstein, and the Search for Reality," in *Thematic Origins of Scientific Thought: Kepler to Einstein*, 2nd ed. (Cambridge, MA: Harvard University Press, 1998), chap. 7; Carl Hoefer, "Einstein's Struggle for a Machian Gravitation Theory," *Studies in History and Philosophy of Science* 25 (1994): 287–335; and Michel Janssen, "Of Pots and Holes: Einstein's Bumpy Road to General Relativity," *Annalen der Physik* 14 Suppl. (2005): 58–85.

6. See, e.g., Laurie Brown, Max Dresden, and Lillian Hoddeson, eds., *Pions to Quarks: Particle Physics in the 1950s* (New York: Cambridge University Press, 1989); Laurie Brown and Helmut Rechenberg, *The Origin of the Concept of Nuclear Forces* (Philadelphia: Institute of Physics Publishing, 1996); and Lillian Hoddeson, Laurie Brown, Michael Riordan, and Max Dresden, eds., *The Rise of the Standard Model: Particle Physics in the 1960s and 1970s* (New York: Cambridge University Press, 1997).

7. Carl H. Brans, "Mach's Principle and a Varying Gravitational Constant" (PhD diss., Princeton University, 1961); Carl H. Brans and Robert H. Dicke, "Mach's Principle and a Relativistic Theory of Gravitation," *Physical Review* 124 (1961): 925–35. On the Caltech group, see Will, *Was Einstein Right?*, 156. Other physicists had introduced similar modifications to general relativity before the Brans-Dicke work, though the earlier efforts had not attracted widespread attention within the community. See Hubert Goenner, "Some Remarks on the Genesis of Scalar-Tensor Theories," *General Relativity and Gravitation* 44 (2012): 2077, https://arxiv.org/abs/1204.3455; and Carl H. Brans, "Varying Newton's Constant: A Personal History of Scalar-Tensor Theories," *Einstein Online* 04 (2010): 1002.

8. Jeffrey Goldstone, "Field Theories with 'Superconductor' Solutions," *Nuovo cimento* 19 (1961): 154–64. See also Laurie Brown and Tian-Yu Cao, "Spontaneous Breakdown of Symmetry: Its Rediscovery and Integration into Quantum Field Theory," *Historical Studies in the Physical and Biological Sciences* 21 (1991): 211–35; and Laurie Brown, Robert Brout, Tian Yu Cao, Peter Higgs, and Yoichiro Nambu, "Panel Session: Spontaneous Breaking of Symmetry," in Hoddeson et al., *Rise of the Standard Model*, 478–522.

9. Peter W. Higgs, "Broken Symmetries, Massless Particles, and Gauge Fields," *Physics Letters B* 12 (1964): 132–33; Peter W. Higgs, "Broken Symmetries and the Masses of Gauge Bosons," *Physical Review Letters* 13 (1964): 508–9; and Peter W. Higgs, "Spontaneous Symmetry Breakdown without Massless Bosons," *Physical Review* 145 (1966): 1156–63.

10. See https://inspirehep.net.

11. These statistics concern citations within the Web of Knowledge database (formerly the Science Citation Index) to the 1961 Brans-Dicke article and to either of Higgs's 1964 articles and/or his 1966 article; during this period, physicists tended to cite some or all the Higgs papers together. I tracked citations using Web of Knowledge rather than the high-energy physics database Inspire because during

the early 1960s, coverage within Inspire tended to focus more narrowly around particle physics rather than gravitation and cosmology. Nonetheless, by October 2018, Inspire included 2,998 citations to the 1961 Brans-Dicke paper, while Higgs's 1964 papers have accumulated 4,893 and 4,192 citations within the Inspire database (respectively), and his 1966 paper has 2,867 citations within Inspire. At that time, Inspire included citation statistics on more than 1 million articles, only 123 of which had been cited 2,998 times or more. See https://inspirehep.net/search?of=hcs&action_search=Search (accessed 24 October 2018).

12. Similarly, although Goldstone's 1961 article on spontaneous symmetry breaking received 487 citations within the Web of Knowledge database between 1961 and 1981, only one paper cited both the Brans-Dicke and Goldstone papers during that period.

13. Physics Survey Committee, *Physics: Survey and Outlook* (Washington, DC: National Academy of Sciences, 1966), 38–45, 52, 95, 111.

14. Cf., e.g., Y. B. Zel'dovich and I. D. Novikov, *Relativistic Astrophysics*, vol. 2, trans. Leslie Fishbone (1975; Chicago: University of Chicago Press, 1983), with Steven Weinberg, *Gravitation and Cosmology* (New York: Wiley, 1972).

15. David Gross and Frank Wilczek, "Ultraviolet Behavior of Nonabelian Gauge Theories," *Physical Review Letters* 30 (1973): 1343–46; David Gross and Frank Wilczek, "Asymptotically Free Gauge Theories, I," *Physical Review D* 8 (1973): 3633–52; David Gross and Frank Wilczek, "Asymptotically Free Gauge Theories, II," *Physical Review D* 9 (1974): 980–93; H. David Politzer, "Reliable Perturbative Results for Strong Interactions?," *Physical Review Letters* 30 (1973): 1346–49; and H. David Politzer, "Asymptotic Freedom: An Approach to Strong Interactions," *Physics Reports* 14 (1974): 129–80.

16. Howard Georgi and Sheldon Glashow, "Unity of All Elementary Particle Forces," *Physical Review Letters* 32 (1974): 438–41. See also Jogesh Pati and Abdus Salam, "Unified Lepton-Hadron Symmetry and a Gauge Theory of the Basic Interactions," *Physical Review D* 8 (1973): 1240–51.

17. See, e.g., Heinz Pagels, *The Cosmic Code: Quantum Physics as the Language of Nature* (New York: Bantam, 1982), 275–77; Paul Davies, *God and the New Physics* (New York: Penguin, 1984), 159–60; John Gribben, *In Search of the Big Bang: Quantum Physics and Cosmology* (New York: Bantam, 1986), 293, 307, 312, 321, 345; Robert Adair, *The Great Design: Particles, Fields, and Creation* (New York: Oxford University Press, 1987), 357; Alan Guth, "Starting the Universe: The Big

Bang and Cosmic Inflation," in *Bubbles, Voids, and Bumps in Time: The New Cosmology*, ed. J. Cornell (New York: Cambridge University Press, 1989), 105–6; Edward W. Kolb, *Blind Watchers of the Sky: The People and Ideas That Shaped Our View of the Universe* (Reading, MA: Addison-Wesley, 1996), 277–80; and Brian Greene, *The Elegant Universe: Superstrings, Hidden Dimensions, and the Quest for the Ultimate Theory* (New York: W. W. Norton, 1999), 177. See also Marcia Bartusiak, *Thursday's Universe: A Report from the Frontier on the Origin, Nature, and Destiny of the Universe* (New York: Times Books, 1986), 227; Timothy Ferris, *Coming of Age in the Milky Way* (New York: Anchor, 1988), 336–37; and Dennis Overbye, *Lonely Hearts of the Cosmos: The Story of the Scientific Quest for the Secret of the Universe* (New York: HarperCollins, 1991), 204, 234.

18. David Schramm, "Cosmology and New Particles," in *Particles and Fields, 1977*, ed. P. A. Schreiner, G. H. Thomas, and A. B. Wicklund (New York: American Institute of Physics, 1978), 87–101; Gary Steigman, "Cosmology Confronts Particle Physics," *Annual Review of Nuclear and Particle Science* 29 (1979): 313–37; and R. J. Tayler, "Cosmology, Astrophysics, and Elementary Particle Physics," *Reports on Progress in Physics* 43 (1980): 253–99. Steigman makes passing reference in his introduction to the new work on grand unification but explicitly labels GUTs as "beyond the scope of this review" (328, 336). Georgi and Glashow's (now-famous) 1974 paper on GUTs ("Unity of All Elementary Particle Forces") received fewer than 50 citations worldwide per year between 1974 and 1978, rapidly rising to more than 200 citations per year beginning in 1980. Anthony Zee likewise recalls that GUTs received little attention, even from particle theorists, until the very end of the 1970s: Anthony Zee, *An Old Man's Toy: Gravity at Work and Play in Einstein's Universe* (New York: Macmillan, 1989), 117.

19. David Kaiser, "Cold War Requisitions, Scientific Manpower, and the Production of American Physicists after World War II," *Historical Studies in the Physical and Biological Sciences* 33 (2002): 131–59; and David Kaiser, "Booms, Busts, and the World of Ideas: Enrollment Pressures and the Challenge of Specialization," *Osiris* 27 (2012): 276–302. See also Daniel Kevles, *The Physicists: The History of a Scientific Community in Modern America*, 3rd ed. (Cambridge, MA: Harvard University Press, 1995), chaps. 24–25.

20. Kevles, *Physicists*, 421.

21. Physics Survey Committee, *Physics in Perspective* (Washington, DC: National Academy of Sciences, 1972), 1:367; and Physics Survey

Committee, *Physics through the 1990s: An Overview* (Washington, DC: National Academy Press, 1986), 98.

22. Physics Survey Committee, *Physics in Perspective*, 1:119.

23. Arthur Beiser to Malcolm Johnson, 14 April 1959, in LIS box 12, folder "Yilmaz: Relativity" ("not a vast market"). Figures on textbook publications come from keyword and call-number searches in the online catalog of the US Library of Congress: http://www.loc.gov. On Feynman's idiosyncratic Caltech course on gravitation, see David Kaiser, "A *Psi* Is Just a *Psi*? Pedagogy, Practice, and the Reconstitution of General Relativity, 1942–1975," *Studies in the History and Philosophy of Modern Physics* 29 (1998): 321–38. See also Achilleus Papetrou, *Lectures on General Relativity* (Boston: Reidel, 1974).

24. Anthony Zee, "Broken-Symmetric Theory of Gravity," *Physical Review Letters* 42 (1979): 417–21; and Lee Smolin, "Towards a Theory of Spacetime Structure at Very Short Distances," *Nuclear Physics B* 160 (1979): 253–68.

25. Yasunori Fujii, "Scalar-Tensor Theory of Gravitation and Spontaneous Breakdown of Scale Invariance," *Physical Review D* 9 (1974): 874–76; F. Englert, E. Gunzig, C. Truffin, and P. Windey, "Conformal Invariant General Relativity with Dynamical Symmetry Breakdown," *Physics Letters B* 57 (1975): 73–77; P. Minkowski, "On the Spontaneous Origin of Newton's Constant," *Physics Letters B* 71 (1977): 419–21; T. Matsuki, "Effects of the Higgs Scalar on Gravity," *Progress of Theoretical Physics* 59 (1978): 235–41; and E. M. Chudnovskii, "Spontaneous Breaking of Conformal Invariance and the Higgs Mechanism," *Theoretical and Mathematical Physics* 35 (1978): 538–39.

26. Zee and Smolin parameterized their gravitational equations slightly differently than Brans and Dicke had done. They followed the usual convention in particle physics of giving scalar fields the dimension of *mass* (for theories defined in four spacetime dimensions). In these units, Newton's gravitational constant G has units $1/(mass)^2$, and hence Zee and Smolin each set G equal to the inverse square of their scalar field rather than to the inverse as in the original Brans-Dicke work.

27. Anthony Zee to John Wheeler, February 1977, included in "Wheeler Family Gathering," vol. 2 (a collection of reminiscences by Wheeler's former students), a copy of which is available in the Niels Bohr Library, call number AR167, American Institute of Physics, College Park, MD; and Anthony Zee, telephone interview with the author, 16 May 2005.

28. Lee Smolin, interview with the author, MIT, 1 December 2004.

See also Lee Smolin, *The Life of the Cosmos* (New York: Oxford University Press, 1997), 7–8, 50; and Lee Smolin, "A Strange Beautiful Girl in a Car," in *Curious Minds: How a Child Becomes a Scientist*, ed. John Brockman (New York: Random House, 2004), 71–78. On Coleman's Harvard course on general relativity, see Kaiser, "A *Psi* Is Just a *Psi*?," 331–33.

29. Edward W. Kolb and Michael S. Turner, *The Early Universe* (Reading, MA: Addison-Wesley, 1990). See also David Kaiser, "Whose Mass Is It Anyway? Particle Cosmology and the Objects of Theory," *Social Studies of Science* 36 (2006): 533–64, on 549–50. On the founding of the Center for Particle Astrophysics at Fermilab, see also Overbye, *Lonely Hearts of the Cosmos*, 206–11; and Steve Nadis, "The Lost Years of Michael Turner," *Astronomy* 32 (April 2004): 44–49, on 48.

30. To be fair, my earlier assumption—like that of the other physicists who had considered models that combined a Brans-Dicke-like gravitational coupling to a Higgs-like field—was that the Higgs field would be associated with some higher-energy symmetry breaking, perhaps at the GUT scale. Hence, I had been focused on different ranges of the various parameters rather than considering the Standard Model Higgs field. See also David Kaiser, "Nonminimal Couplings in the Early Universe: Multifield Models of Inflation and the Latest Observations," in *At the Frontier of Spacetime: Scalar-Tensor Theory, Bell's Inequality, Mach's Principle, Exotic Smoothness*, ed. T. Asselmeyer-Maluga (New York: Springer, 2016), 41–57, http://arxiv.org/abs/1511.09148.

Chapter 14

A version of this essay originally appeared in *London Review of Books* 32 (8 July 2010): 34–35.

1. Ki Mae Heussner, "Stephen Hawking: Alien Contact Could Be Risky," 26 April 2010, ABCNews.com.

2. Steven J. Dick, *Life on Other Worlds: The Twentieth Century Extraterrestrial Life Debate* (New York: Cambridge University Press, 2001).

3. Giuseppe Cocconi and Philip Morrison, "Searching for Interstellar Communications," *Nature* 184 (19 September 1959): 844–46.

4. Cocconi and Morrison, "Searching for Interstellar Communications."

5. Cocconi and Morrison, "Searching for Interstellar Communications."

6. Silvan S. Schweber, *In the Shadow of the Bomb: Oppenheimer,*

Bethe, and the Moral Responsibility of the Scientist (Princeton: Princeton University Press, 2000), 130–45.

7. Frank Drake and Dava Sobel, *Is Anyone Out There? The Scientific Search for Extraterrestrial Intelligence* (New York: Delacorte, 1992).

8. Paul Davies, *The Eerie Silence: Renewing Our Search for Alien Intelligence* (Boston: Houghton Mifflin, 2010).

9. Davies, *Eerie Silence*, 175.

10. Jennifer Burney, "The Search for Extraterrestrial Intelligence: Changing Science Here on Earth" (AB thesis, Harvard University, 1999).

11. Burney, "Search for Extraterrestrial Intelligence," 78–84. On more recent efforts, see, e.g., Chelsea Gohd, "Breakthrough Listen Launches New Search for E.T. across Millions of Stars," 8 May 2018, Space.com.

12. Burney, "Search for Extraterrestrial Intelligence," chap. 4.

13. Davies, *Eerie Silence*, 198.

14. Peter Galison and Robb Moss, *Containment* (documentary film, 2015), http://www.containmentmovie.com.

Chapter 15

A version of this essay originally appeared in *Isis* 103 (March 2012): 126–38.

1. Charles W. Misner, Kip S. Thorne, and John A. Wheeler, *Gravitation* (San Francisco: W. H. Freeman, 1973). On nicknames for the book, see, e.g., "Chicago Undergraduate Physics Bibliography," accessed 8 July 2011, http://www.ocf.berkeley.edu/~abhishek/chicphys.htm.

2. For succinct introductions to the early history of Einstein's work on general relativity, see Michel Janssen, "'No Success like Failure': Einstein's Quest for General Relativity," in *The Cambridge Companion to Einstein*, ed. Michel Janssen and Christoph Lehner (New York: Cambridge University Press, 2014), 167–227; Hanoch Gutfreund and Jürgen Renn, *The Road to Relativity: The History and Meaning of Einstein's "The Foundation of General Relativity"* (Princeton: Princeton University Press, 2015); Michel Janssen and Jürgen Renn, "Arch and Scaffold: How Einstein Found His Field Equations," *Physics Today* 68, no. 11 (November 2015): 30–36; and Matthew Stanley, *Einstein's War: How Relativity Triumphed amid the Vicious Nationalism of World War I* (New York: Dutton, 2019).

3. Albert Einstein, foreword to Peter G. Bergmann, *Introduction to the Theory of Relativity* (New York: Prentice-Hall, 1942), v. On Edding-

ton's eclipse expedition and the early reception of general relativity, see Jean Eisenstaedt, *The Curious History of Relativity: How Einstein's Theory of Gravity Was Lost and Found Again* (Princeton: Princeton University Press, 2006); Jeffrey Crelinstein, *Einstein's Jury: The Race to Test Relativity* (Princeton: Princeton University Press, 2006); Matthew Stanley, *Practical Mystic: Religion, Science, and A. S. Eddington* (Chicago: University of Chicago Press, 2007), chap. 3; Hanoch Gutfreund and Jürgen Renn, *The Formative Years of Relativity: The History and Meaning of Einstein's Princeton Lectures* (Princeton: Princeton University Press, 2017); Daniel Kennefick, *No Shadow of a Doubt: The 1919 Eclipse That Confirmed Einstein's Theory of Relativity* (Princeton: Princeton University Press, 2019); and Stanley, *Einstein's War*.

4. On the return of general relativity to physics departments' course offerings during the 1950s and 1960s, see David Kaiser, "A *Psi* Is Just a *Psi*? Pedagogy, Practice, and the Reconstitution of General Relativity, 1942–1975," *Studies in the History and Philosophy of Modern Physics* 29 (1998): 321–38; Daniel Kennefick, *Traveling at the Speed of Thought: Einstein and the Quest for Gravitational Waves* (Princeton: Princeton University Press, 2007), chap. 6; and Alexander Blum, Roberto Lalli, and Jürgen Renn, "The Reinvention of General Relativity: A Historiographical Framework for Assessing One Hundred Years of Curved Space-Time," *Isis* 106 (September 2015): 598–620. On Wheeler as an effective mentor, see Charles W. Misner, Kip S. Thorne, and Wojciech H. Zurek, "John Wheeler, Relativity, and Quantum Information," *Physics Today* 62, no. 4 (April 2009): 40–46; and Terry M. Christensen, "John Wheeler's Mentorship: An Enduring Legacy," *Physics Today* 62, no. 4 (April 2009): 55–59.

5. Steven Weinberg, *Gravitation and Cosmology: Principles and Applications of the General Theory of Relativity* (New York: Wiley, 1972); and S. W. Hawking and G. F. R. Ellis, *The Large Scale Structure of Space-Time* (New York: Cambridge University Press, 1973).

6. John Wheeler, handwritten notes, "Thoughts on preface, Mon., 13 July 1970," in JAW series IV, box F-L, folder "Gravitation: Notes with Charles W. Misner and Kip S. Thorne" ("committee planning graduate courses"). See also form letter from Misner, Thorne, and Wheeler to colleagues announcing forthcoming publication of the book, 13 June 1973, in KST folder "MTW: Sample pages."

7. John Wheeler, handwritten notes, page for insertion into draft of preface, n.d. (late August 1970) ("third channel of pedagogy"), and Wheeler, handwritten notes, "Plan of Book, Sat., 18 July 1970" ("*test a write up*" [emphasis in original]), both in JAW series IV, box F-L,

folder "Gravitation: Notes with Charles W. Misner and Kip S. Thorne."
On sidebars in more elementary physics textbooks, see Sharon Tra-
week, *Beamtimes and Lifetimes: The World of High-Energy Physicists*
(Cambridge, MA: Harvard University Press, 1988), 76–81.

8. Kip Thorne to Earl Tondreau (editor at W. H. Freeman), 14 October 1970, in KST folder "MTW: Correspondence, 1970–May, 1973"
("several features," typefaces). See also Thorne to Robert Ishikawa and
Aidan Kelley (W. H. Freeman), 28 January 1971, in KST folder "MTW";
and Evan Gillespie (W. H. Freeman) to Kip Thorne, 29 November 1972,
in KST folder "MTW: Publishing company, 1970–71, 1971–72."

9. Kip Thorne to Y. B. Zel'dovich and I. D. Novikov, 21 June 1973, in
KST folder "MTW: Correspondence, June, 1973–."

10. Thorne to Ishikawa and Kelley, 28 January 1971 ("dependency
statements").

11. Kip Thorne to John Wheeler and Charles Misner, with cc to
Bruce Armbruster, 17 February 1972, in KST folder "MTW: Correspondence, 1970–May, 1973."

12. Thorne to Wheeler and Misner with cc to Armbruster, 17 February 1972. See also Misner, Thorne, and Wheeler, form letter to colleagues, 13 June 1973.

13. Thorne to Bruce Armbruster, 10 April 1973 (royalty rates,
pricing vis-à-vis Weinberg's book, "capture one hundred percent"),
in KST folder "MTW: Publishing company, 1970–71, 1971–72." On
pricing, see also Thorne to Richard Warrington (president), Peter
Renz (science editor), and Lew Kimmick (financial manager) at
W. H. Freeman, 14 February 1979, in JAW series II, box Fr-Gl, folder
"W. H. Freeman and Co., Publishers"; Thorne to Wheeler and Misner,
2 November 1972, in KST folder "MTW"; Misner to Wheeler and
Thorne, 18 November 1982, in KST folder "MTW" (copy also in JAW
series II, box Fr-Gl, folder "W. H. Freeman and Co., Publishers"); and
royalty statement from June 1993, in KST folder "MTW: Royalty
statements."

14. Dennis Sciama, "Modern View of General Relativity," *Science*
183 (22 March 1974): 1186 ("pedagogic masterpiece"); Michael Berry,
review in *Science Progress* 62, no. 246 (1975): 356–60, on 360 ("Aladdin's cave"); and David Park, "Ups and Downs of 'Gravitation,'" *Washington Post*, 21 April 1974, 4 ("three highly inventive people"). See
also D. Allan Bromley, review in *American Scientist* (January–February
1974): 101–2.

15. L. Resnick, review in *Physics in Canada*, June 1975, clipping
in KST folder "MTW: Reviews" ("difficult book to read"); S. Chandra-

sekhar, "A Vast Treatise on General Relativity," *Physics Today*, August 1974, 47–48, on 48 ("needless repetition"); and W. H. McCrea, review in *Contemporary Physics* 15, no. 4 (July 1974), clipping in KST folder "MTW: Reviews" ("variety of gimmicks").

16. John Wheeler, handwritten notes, "Thoughts on preface, Mon., 13 July 1970" ("make clear the idea"). On Wheeler's style, see also John A. Wheeler with Kenneth Ford, *Geons, Black Holes, and Quantum Foam: A Life in Physics* (New York: W. W. Norton, 1998); and Misner, Thorne, and Zurek, "John Wheeler, Relativity, and Quantum Information."

17. Sciama, "Modern View of General Relativity," 1186 ("prose style"); Resnick, review in *Physics in Canada* ("commendable attempt"); and J. Bicak, review in *Bulletin of the Astronomical Institute of Czechoslovakia* 26, no. 6 (1975): 377–78 ("A 'poetical' style").

18. Alan Farmer, review in *Journal of the British Interplanetary Society* 27 (1974): 314–15, on 314 ("comes dangerously close"); and Ian Roxburgh, "Geometry Is All, or Is It?," *New Scientist*, 26 September 1974, 828 ("a regular subscriber").

19. Chandrasekhar, "A Vast Treatise on General Relativity," 48; and Thorne to Chandrasekhar, 21 June 1974, in KST folder "MTW: Reviews." On Chandrasekhar's career, see K. C. Wali, *Chandra: A Biography of S. Chandrasekhar* (Chicago: University of Chicago Press, 1991); and Arthur I. Miller, *Empire of the Stars: Obsession, Friendship, and Betrayal in the Quest for Black Holes* (Boston: Houghton Mifflin, 2005).

20. Kip Thorne to Peter Renz, 15 June 1983, in KST folder "MTW" ("large fraction"); and Thorne to Warrington, Renz, and Kimmick, 14 February 1979, on annual sales of *Gravitation* and Weinberg's textbook.

21. Sales figures from royalty statement of June 1993 in KST folder "MTW: Royalty statements." On PhD conferral rates, see David Kaiser, "Cold War Requisitions, Scientific Manpower, and the Production of American Physicists after World War II," *Historical Studies in the Physical and Biological Sciences* 33 (2002): 131–59; and David Kaiser, "Booms, Busts, and the World of Ideas: Enrollment Pressures and the Challenge of Specialization," *Osiris* 27 (2012): 276–302.

22. Kip Thorne to Peter Renz, 10 August 1983, in KST folder "MTW."

23. Park, "Ups and Downs of 'Gravitation,'" 4.

24. Robert Pincus, "Gravity Theory Excites the Mind," clipping in KST folder "MTW: Reviews." The clipping does not indicate date, pub-

lication title, or page number, but advertisements on the same page as the review clearly indicate that the newspaper was based in San Antonio, Texas.

25. See, e.g., Andrzej Trautman to Charles Misner, Kip Thorne, and John Wheeler, 10 January 1974, in KST folder "MTW"; Heinz Pagels to Wheeler, 1 February 1974, in KST folder "MTW Reviews"; Philip B. Burt to Wheeler, 12 November 1974, in KST folder "MTW"; and Robert Rabinoff to Misner, Thorne, and Wheeler, 10 March 1978, in KST folder "MTW: Reviews."

26. Luigi Vignato to Charles Misner, Kip Thorne, and John Wheeler, 20 July 1976, in KST folder "MTW: Correspondence, June, 1973–"; and Wheeler to Vignato, 2 August 1976, in the same folder. Wheeler did not directly address Vignato's question, but he did enclose a preprint of his recent essay: John Wheeler, "Genesis and Observership," in *Foundational Problems in the Special Sciences*, ed. Robert E. Butts and Jaakko Hintikka (Boston: Reidel, 1977), 3–33.

27. Jadoul Michel to Charles Misner, Kip Thorne, and John Wheeler, August 1983, in KST folder "MTW."

28. Dan Foley to Kip Thorne, 7 February 1980, in KST folder "MTW." See also Thorne to Foley, 27 February 1980, in the same folder.

29. John Wheeler to Peter Renz, 28 June 1979, in KST folder "MTW"; copy also in JAW series II, box Fr-Gl, folder "W. H. Freeman and Co. Publishers."

30. Charles W. Misner, Kip S. Thorne, and John A. Wheeler, *Gravitation* (repr., Princeton: Princeton University Press, 2017).

Chapter 16

A version of this essay originally appeared in *American Scientist* 95 (November–December 2007): 518–25.

1. On various fragments attributed to Heraclitus, especially regarding the nature of change, see Daniel W. Graham, "Heraclitus," in *Stanford Encyclopedia of Philosophy* (Fall 2015 ed.), sec. 3.1 ("flux"), https://plato.stanford.edu/archives/fall2015/entries/heraclitus.

2. Christopher Smeenk, "Einstein's Role in the Creation of Relativistic Cosmology," in *The Cambridge Companion to Einstein*, ed. Michel Janssen and Christoph Lehner (New York: Cambridge University Press, 2014), 228–69.

3. Helge Kragh, *Cosmology and Controversy: The Historical Development of Two Theories of the Universe* (Princeton: Princeton University Press, 1996), chap. 2; and Eduard Tropp, Viktor Y. Frenkel, and Artur

Chernin, *Alexander A. Friedmann: The Man Who Made the Universe Expand* (New York: Cambridge University Press, 2006).

4. Kragh, *Cosmology and Controversy*, chap. 2. See also Dominique Lambert, *The Atom of the Universe: The Life and Work of Georges Lemaître*, trans. Luc Ampleman (New York: Copernicus Center Press, 2015); and Helge Kragh and Robert W. Smith, "Who Discovered the Expanding Universe?," *History of Science* 41 (2003): 141–62. On Lemaître, Hubble, and the early interpretations of Hubble's data in terms of cosmic expansion, see also Mario Livio, "Mystery of the Missing Text Solved," *Nature* 479 (10 November 2011): 171–73; and Elizabeth Gibney, "Belgian Priest Recognized in Hubble-Law Name Change," *Nature*, 30 October 2018, https://www.nature.com/articles/d41586-018-07234-y.

5. Quoted in Kragh, *Cosmology and Controversy*, 48–49.

6. Quoted in Kragh, *Cosmology and Controversy*, 46 (Eddington), 56 (Tolman). On Eddington's approach, see also Matthew Stanley, *Practical Mystic: Religion, Science, and A. S. Eddington* (Chicago: University of Chicago Press, 2007).

7. Edward Larson, *Summer for the Gods: The Scopes Trial and America's Continuing Debate over Science and Religion* (New York: Basic, 1997); and Adam Shapiro, *Trying Biology: The Scopes Trial, Textbooks, and the Antievolution Movement in American Schools* (Chicago: University of Chicago Press, 2013).

8. "Topics of the Times," *New York Times*, 6 February 1923, 18; Simeon Strunsky, "About Books, *More or Less*: Excessively Up to Date," *New York Times*, 29 April 1928, BR3; "By-Products: In the Matter of Einstein, Tea-Kettles, Destiny, &c.," *New York Times*, 22 March 1931, E1; and "Improving on Relativity," *New York Times*, 15 March 1939, 18.

9. On the Hopkins Applied Physics Laboratory, see Michael Aaron Dennis, "'Our First Line of Defense': Two University Laboratories in the Postwar American State," *Isis* 85 (1994): 427–55.

10. Kragh, *Cosmology and Controversy*, chap. 3.

11. George Gamow, *The Creation of the Universe* (New York: Viking, 1952); and George Gamow, "The Role of Turbulence in the Evolution of the Universe," *Physical Review* 86 (1952): 251.

12. Kragh, *Cosmology and Controversy*, chap. 4.

13. Fred Hoyle, *The Nature of the Universe* (New York: Harper, 1950). See also Helge Kragh, "Naming the Big Bang," *Historical Studies in the Natural Sciences* 44 (2012): 3–36.

14. Ronald Numbers, *The Creationists* (New York: Knopf, 1992), chap. 10.

15. Kragh, *Cosmology and Controversy*, chap. 7; and Steven Weinberg, *The First Three Minutes* (New York: Basic, 1977).

16. Dean Rickles, *A Brief History of String Theory* (New York: Springer, 2014). See also Brian Greene, *The Elegant Universe: Superstrings, Hidden Dimensions, and the Quest for the Ultimate Theory* (New York: W. W. Norton, 1999).

17. Lee Smolin, *The Trouble with Physics: The Rise of String Theory, the Fall of a Science, and What Comes Next* (Boston: Houghton Mifflin, 2006). See also Peter Woit, *The Failure of String Theory and the Search for Unity in Physical Law* (New York: Basic, 2006).

18. See, e.g., Lisa Randall, *Warped Passages: Unraveling the Mysteries of the Universe's Hidden Dimensions* (New York: Ecco, 2005).

19. Leonard Susskind, *The Cosmic Landscape: String Theory and the Illusion of Intelligent Design* (New York: Little, Brown, 2005).

20. "Billionaires: The Richest People in the World," *Forbes*, 5 March 2019, https://www.forbes.com/billionaires/#3e3f70c1251c.

21. For an accessible introduction to inflationary cosmology, see Alan Guth, *The Inflationary Universe: The Quest for a New Theory of Cosmic Origins* (New York: Basic, 1997).

22. Susskind, *Cosmic Landscape*; and Alexander Vilenkin, *Many Worlds in One: The Search for Other Universes* (New York: Farrar, Straus, and Giroux, 2007). On earlier discussions of the "anthropic principle" in physics, see John Barrow and Frank Tipler, eds., *The Anthropic Cosmological Principle* (New York: Oxford University Press, 1986).

23. Bernard le Bovier de Fontenelle, *Conversations on the Plurality of Worlds* (1686), trans. H. A. Hargreaves (Berkeley: University of California Press, 1990); and Isaac Newton, *Four Letters from Sir Isaac Newton to Doctor Bentley, Containing Some Arguments in Proof of a Deity* (London: R. and J. Dodsley, 1756). See also Rob Iliffe, "The Religion of Isaac Newton," in *The Cambridge Companion to Newton*, 2nd ed. (New York: Cambridge University Press, 2016), 485–523.

24. Susskind, *Cosmic Landscape*, vii.

25. See, e.g., Dennis Overbye, "Zillions of Universes? Or Did Ours Get Lucky?," *New York Times*, 28 October 2003.

26. Bacon quoted in James Glanz, "Science vs. the Bible: Debate Moves to the Cosmos," *New York Times*, 10 October 1999.

27. Laurie Goodstein, "Judge Rejects Teaching Intelligent Design," *New York Times*, 21 December 2005; and Andrew Revkin, "A Young Bush Appointee Resigns His Post at NASA," *New York Times*, 8 February 2006. See also John Brockman, ed., *Intelligent Thought: Science versus the Intelligent Design Movement* (New York: Vintage, 2006).

28. Numbers, *Creationists*, chap. 9.

29. David F. Coppedge, "State of the Cosmos Address Offered," 21 February 2005, https://crev.info/2005/02/state_of_the_cosmos _address_offered. Cf. Alan Guth and David Kaiser, "Inflationary Cosmology: Exploring the Universe from the Smallest to the Largest Scales," *Science* 307 (11 February 2005): 884–90, https://arxiv.org/abs /astro-ph/0502328.

Chapter 17

A version of this essay originally appeared in *London Review of Books* 33 (17 February 2011): 36–37.

1. Dennis Overbye, *Lonely Hearts of the Cosmos: The Scientific Quest for the Secret of the Universe* (New York: HarperCollins, 1991).

2. On the *COBE* mission, see, e.g., George Smoot with Keay Davidson, *Wrinkles in Time* (New York: William Morrow, 1993).

3. The numerical values quoted here come from N. Aghanim et al. (*Planck* Collaboration), "*Planck* 2018 Results, VI: Cosmological Parameters," http://arxiv.org/abs/1807.06209. After the *Planck* team released its 2015 measurements, other groups, using observations of distinct astrophysical phenomena (such as supernovae), have measured a value for the Hubble expansion rate that differs by about 8 percent from the *Planck* value. Whether the distinct measurements will eventually converge or whether the modest discrepancy points to some new, unexplained physics remains to be seen. See Joshua Sokol, "Hubble Trouble," *Science*, 10 March 2017, 1010–14.

4. For an accessible introduction, see, e.g., David Weintraub, *How Old Is the Universe?* (Princeton: Princeton University Press, 2010).

5. Penrose describes much of this work in Roger Penrose, *Cycles of Time* (New York: Knopf, 2010).

6. Aaron Wright, "The Origins of Penrose Diagrams in Physics, Art, and the Psychology of Perception, 1958–1962," *Endeavor* 37, no. 3 (2013): 133–39. See also Aaron Wright, "The Advantages of Bringing Infinity to a Finite Place: Penrose Diagrams as Objects of Intuition," *Historical Studies in the Natural Sciences* 44, no. 2 (2014): 99–139.

7. Lisa Randall, *Warped Passages: Unraveling the Mysteries of the Universe's Hidden Dimensions* (New York: Ecco, 2005).

8. V. G. Gurzadyan and R. Penrose, "Concentric Circles in WMAP Data May Provide Evidence of Violent Pre-Big-Bang Activity," http:// arxiv.org/abs/1011.3706; and V. G. Gurzadyan and R. Penrose, "More

on the Low Variance Circles in CMB Sky," http://arxiv.org/abs/1012
.1486. Penrose has continued to investigate these ideas: V. G. Gurzadyan and R. Penrose, "CCC-Predicted Low-Variance Circles in CMB Sky
and LCDM," http://arxiv.org/abs/1104.5675; V. G. Gurzadyan and
R. Penrose, "On CCC-Predicted Concentric Low-Variance Circles in the
CMB Sky," *European Physical Journal Plus* 128 (2013): 22, http://arxiv
.org/abs/1302.5162; V. G. Gurzadyan and R. Penrose, "CCC and the
Fermi Paradox," *European Physical Journal Plus* 131 (2016): 11, http://
arxiv.org/abs/1512.00554; and Roger Penrose, "Correlated 'Noise' in
LIGO Gravitational Wave Signals: An Implication of Conformal Cyclic
Cosmology," http://arxiv.org/abs/1707.04169. For the early responses
that found no support for Penrose's model within the *WMAP* data,
see Adam Moss, Douglas Scott, and James Zibin, "No Evidence for
Anomalously Low Variance Circles on the Sky," *Journal of Cosmology
and Astro-Particle Physics* 1104 (2011): 033, http://arxiv.org/abs/1012
.1305; I. K. Wehus and H. K. Eriksen, "A Search for Concentric Circles
in the 7-Year WMAP Temperature Sky Maps," *Astrophysical Journal*
733 (2011): L29, http://arxiv.org/abs/1012.1268; and Amir Hajian,
"Are There Echoes from the Pre–Big Bang Universe? A Search for Low
Variance Circles in the CMB Sky," *Astrophysical Journal* 740 (2011): 52,
http://arxiv.org/abs/1012.1656.

Chapter 18

A version of this essay originally appeared in *New York Times*, 3 October 2017.

1. B. P. Abbott et al. (LIGO Scientific Collaboration and Virgo Collaboration), "Observation of Gravitational Waves from a Binary Black
Hole Merger," *Physical Review Letters* 116 (2016): 061102, http://arxiv
.org/abs/1602.03837. See also Janna Levin, *Black Hole Blues, and
Other Songs from Outer Space* (New York: Knopf, 2016); Stefan Helmreich, "Gravity's Reverb: Listening to Space-Time, or Articulating
the Sounds of Gravitational-Wave Detection," *Cultural Anthropology*
31 (2016): 464–92; and Harry Collins, *Gravity's Kiss: The Detection of
Gravitational Waves* (Cambridge, MA: MIT Press, 2017).

2. Daniel Kennefick, *Traveling at the Speed of Thought: Einstein and
the Quest for Gravitational Waves* (Princeton: Princeton University
Press, 2007).

3. See esp. Harry Collins, *Gravity's Shadow: The Search for Gravitational Waves* (Chicago: University of Chicago Press, 2004), pt. 1.

4. Collins, *Gravity's Shadow*, chap. 17.

5. Charles W. Misner, Kip S. Thorne, and John A. Wheeler, *Gravitation* (San Francisco: W. H. Freeman, 1973), 1014–18.

6. Weiss's proposals and interim progress reports to the National Science Foundation, as quoted in Collins, *Gravity's Shadow*, 280 ("Gravitation research"), 287 ("slowly come to the realization").

7. Collins, *Gravity's Shadow*, pt. 4. On the complicated process of selecting sites for the LIGO project, see also Tiffany Nichols, "Constructing Stillness: The Site Selection History and Signal Epistemological Development of the Laser Interferometer Gravitational-Wave Observatory (LIGO)" (PhD diss., Harvard University, in preparation).

8. Based on data in the ProQuest "Dissertations and Theses" database, with keyword searches for "LIGO" in titles and abstracts.

9. Committee on Accuracy of Time Transfer in Satellite Systems, Air Force Studies Board, *Accuracy of Time Transfer in Satellite Systems* (Washington, DC: National Academy Press, 1986).

Chapter 19

A version of this essay originally appeared in *New Yorker*, 15 March 2018 (online).

1. Stephen Hawking, *A Brief History of Time* (New York: Bantam, 1988). On publishing trends in popular physics around that time, see Elizabeth Leane, *Reading Popular Physics: Disciplinary Skirmishes and Textual Strategies* (London: Ashgate, 2007).

2. Hélène Mialet, *Hawking Incorporated: Stephen Hawking and the Anthropology of the Knowing Subject* (Chicago: University of Chicago Press, 2012).

3. Alan Guth et al., "A Cosmic Controversy," *Scientific American*, July 2017, 5–7.

4. The short film is available at https://www.youtube.com/watch?v=Hi0BzqV_b44.

INDEX

Page numbers in italics refer to figures.

Farmelo, Graham, 25, 27–28
fascism, 36, 41, 45, 89
federal spending on particle
    physics, 195
Fermi, Enrico, 41, 126; affected
    by Nazi-inspired racial laws
    in Italy, 41; Manhattan Proj-
    ect, 41, 42; neutrino, 41;
    Noble Prize, 41; radioactive
    decay, 41, 44; in United
    States, 41
Fermilab accelerator, 161
fermions, 175
Feshbach, Herman, 131
Feynman, Richard, 116, *118*, 119,
    126, 134, 174, 197; curricu-
    lum for physics students,
    116–17, 119; *Feynman Lec-
    tures on Physics, The*, 117, 118;
    "Physics X" course, 134, *135*
Finnegans Wake (Joyce), 165
First World War ("a chemists'
    war"), 72
fission bombs, 91
Fitch, Val, 8, *10*
"five sigma," 180
Flexner, Abraham, 89
Fontenelle, Bernard Le Bovier de,
    244
Formaggio, Joseph, 49, 50
Fowler, Ralph, 18, 19
freedom-of-choice loophole, 57–
    58, 60, 63–64, 67
French Revolution, 88
Friedman, Andrew, 58, 59, *60*,
    64, 236
Friedmann, Alexander, 232–33,
    245
Fuchs, Klaus, 24, 46; atomic
    secrets to the Soviets, 46
Fuld, Caroline Bamberger, 89

funding for research, 82–83;
    Atomic Energy Commission,
    83; Department of Defense,
    83; federal defense agen-
    cies, 83

Gaia hypothesis, 215
Galileo, 235
Galison, Peter, 216
Gallicchio, Jason, 58, *60*, 64
Gamow, George, 236, *237*, 238
Gell-Mann, Murray, 165, *166*,
    173; Noble Prize, 165
*General Account of Methods of
    Using Atomic Energy for Mili-
    tary Purposes under the Aus-
    pices of the United States
    Government, 1940–1945, A*
    (Smyth report), 79–80
general relativity, 185–86, 188,
    197, 199, 218–19, 231–32,
    236, 238, 241, 253, 257, 264,
    269, 271–72
Genesis Flood, The (Whitcomb
    and Morris), 240
Gerjuoy, Edward, 129, 130
Gerstenstein, Michaeil, 266
Gianotti, Fabiola, *181*
G.I. Bill, 83
gluons, 172–73
Gödel, Kurt, 89
God particle, 175
Gold, Thomas, 238, 239
Goldstone, Jeffrey, 189, *190*
Goldstone-Higgs symmetry-
    breaking potential, 197
Goudsmit, Samuel, 9–10
graduate enrollments in physics,
    83–84, 97, 108–10, 112–13,
    120, 124–25, 133–34, 137,
    149, 161, 195

grand unified theories (GUT), 193–95

Gravitation (Misner, Thorne, and Wheeler), 218–30; typesetting, 222–23, 226

gravitational waves, 263–69; detector, *265*

Gravitation and Cosmology (Weinberg), 199, 220, 224–26; reviews, 224–25

gravity, 186; Newton's constant (*G*), 186. *See also* general relativity

Gravity's Shadow: The Search for Gravitational Waves (Collins), 267

Grier, David Alan, 90

Gross, David, 193, 245

Group Portrait with Lady, 17

Groves, Leslie, 79, 82

Guggenheim Foundation, 95

Guralnik, Gerald, 177

GUT. *See* grand unified theories (GUT)

Guth, Alan, 248, 249

Haan, Bierens de, 89

Hagen, Carl, 177

Half-Life (Close), 45

Handsteiner, Johannes, 61; Cosmic Bell test, 62; telescope, 61

Hartnett, John, 246

Hawking, Stephen, 7, 205, 220, 270, *273*; amyotrophic lateral sclerosis (ALS), 272; on black holes, 272–73; on Einstein's equations, 271

Hawking Incorporated (Mialet), 274

Heisenberg, Werner, 2, 17, 18, 19, *21*, 22, *24*, 30, 33, 121; approach to quantum mechanics, 20; discrete arrays of numbers, 20; electrons' paths, 19; first-principles treatment of matter and radiation, 19; uncertainty principle, 2, 119, 122, 172

Heisenberg uncertainty principle, 2, 119, 122, 172

Heraclitus, 231

Herbert, Nick, 150

Herman, Robert, 236, *237*

Higgs, Peter, 177, *181*, 191–92; Noble Prize, 181

Higgs boson, 13, 167, 174, 177, 180–82; decaying, *179*; mass, 178

Higgs fields, 172, 197

Higgs Hunter's Guide, The, 179

Higgs inflation, 183, 201

Hiroshima and Nagasaki, 42, 71, 78

Hitler, Adolf, 3, 31, 34

Hochrainer, Armin, *66*

holism, 146

House Un-American Activities Committee, 81

Hoyle, Fred, 238, 239

Hubble, Edwin, 233

Hubble Space Telescope, 252

Humphreys, Russell, 246

hydrogen atom, 251

hydrogen bombs, 91; calculations for, 93; crash-course development, 94; tritium for, 43

hype-amplification-feedback process, 99

IceCube Neutrino Observatory, 52

early life and schooling, 120–21; teaching at Berkeley and Caltech, 121
organicism, 146
Origin of Species (Darwin), 248
Overbye, Dennis, 250

particle accelerator, Geneva, 155. *See also* Large Hadron Collider (LHC)
particle cosmology, 184, 195
particle physics, 174; crisis, 1970s, 185, 196
Pauli, Wolfgang, 2, 18, 22, 41
Pearl Harbor attack, 73
Penrose, Roger, 255–62, 271; conformal cyclic cosmology (CCC), 256
Penrose diagrams, 257–58
Penrose-Hawking singularity theorems, 271
"phone book, the." See *Gravitation* (Misner, Thorne, and Wheeler)
photons, 254; electric field, 56; of light, 56; massless, 175
Physical Review (Journal), 8, 9, 11–12
"physicists' war," 12, 72–73, 78; Google n-gram of, 79
"Physicist's War, A" (bulletins), 72–73, 75
Physics Survey Committee, 192, 196
Physics: Survey and Outlook, 192
"Physics X," 134
Pius XII, 238
Placement Service of the American Institute of Physics, 110
Planck satellite, 253, 254
Plato, 231

plutonium-producing nuclear reactors, 81
Podolsky, Boris, 32
polarization, 56
Polchinski, Joseph, 242
Politzer, H. David, 193
Pontecorvo, Bruno, 44–45, 47; British nuclear research facility in Harwell, 45; in Fermi's group ("Puppy"), 44–46; Manhattan Project, 45; moved to New York City, 45; on neutrinos, 47–48; nuclear fission, 45; nuclear reactor, 45; nuclear research facility at Dubna, 45; patent dispute, 46; into Soviet territory, 46; theory of neutrino oscillations, 50
Pontecorvo Affair, The (Turchetti), 45
Pontecorvo's theory of neutrino oscillations, 50
Portrait of Isaac Newton, A (Manuel), 26
positron, 43
primeval atom, 234
Principles of Quantum Mechanics, The (Dirac), 22
Privileged Planet, The (DVD), 249
Project Ozma, 206
Prony, Gaspard Riche de, 88
protons, 172
prussic acid, 35
ψ-function, 33, 35
Pustovoit, V. I., 266

quantum chess, 275
quantum electrodynamics (QED), 22